John Lubbock

Fifty Years of Science

Being the address delivered at York to the British Association, August 1881

John Lubbock

Fifty Years of Science
Being the address delivered at York to the British Association, August 1881

ISBN/EAN: 9783337034535

Printed in Europe, USA, Canada, Australia, Japan

Cover: Foto ©ninafisch / pixelio.de

More available books at **www.hansebooks.com**

FIFTY YEARS OF SCIENCE

BEING THE ADDRESS DELIVERED AT YORK

TO THE BRITISH ASSOCIATION

AUGUST 1881

BY

SIR JOHN LUBBOCK, BART., M.P.

PRESIDENT OF THE ASSOCIATION

London

MACMILLAN AND CO.

1882

LONDON : PRINTED BY
SPOTTISWOODE AND CO., NEW-STREET SQUARE
AND PARLIAMENT STREET

ADDRESS

TO THE

BRITISH ASSOCIATION, 1881.

In the name of the British Association, which for the time I very unworthily represent, I beg to tender to you, my Lord Mayor, and through you to the City of York, our cordial thanks for your hospitable invitation and hearty welcome.

We feel, indeed, that in coming to York we are coming home. Gratefully as we acknowledge, and much as we appreciate the kindness we have experienced elsewhere, and the friendly relations which exist between this Association and most—I might even say, all—our great cities, yet Sir R. Murchison truly observed at the close of our first meeting in 1831, that to York, 'as the cradle of the Association, we shall ever look back with gratitude; and whether we meet hereafter on the banks of the Isis, the Cam, or the Forth, to this spot we shall still fondly revert.' Indeed, it would have been a matter of much regret to all of us, if we had not been able on this, our fiftieth anniversary, to hold our meeting in our mother city.

My Lord Mayor, before going further, I must express my regret, especially when I call to mind the illustrious men who have preceded me in this chair, that it has not fallen to one of my eminent friends around me, to preside on this auspicious occasion. Conscious, however, as I am of my own deficiencies, I feel that I must not waste time in dwelling on them, more especially as in doing so I should but give them greater prominence. I will, therefore, only make one earnest appeal to your kind indulgence.

The connection of the British Association with the City of York does not depend merely on the fact that our first meeting was held here. It originated in a letter addressed by Sir D. Brewster to Professor Phillips, as Secretary to your York Philosophical Society, by whom the idea was warmly taken up. The first meeting was held on September 26, 1831, the chair being occupied by Lord Milton, who delivered an address, after which Mr. William Vernon Harcourt, Chairman of the Committee of Management, submitted to the meeting a code of rules which had been so maturely considered, and so wisely framed, that they have remained substantially the same down to the present day.

Of those who organised and took part in that first meeting, few, alas! remain. Brewster and Phillips, Harcourt and Lord Milton, Lyell and Murchison, all have passed away, but their memories live among us. Some few, indeed, of those present at our first meeting, we rejoice to see here to-day, including one of the five members constituting the original organising Committee, our venerable Vice-President, Archdeacon Creyke.

The constitution and objects of the Association were

so ably described by Mr. Spottiswoode, at Dublin, and are so well known to you, that I will not dwell on them this evening. The excellent President of the Royal Society, in the same address, suggested that the past history of the Association would form an appropriate theme for the present meeting. The history of the Association, however, is really the history of science, and I long shrank from the attempt to give even a panoramic survey of a subject so vast and so difficult; nor should I have ventured to make any such attempt, but that I knew I could rely on the assistance of friends in every department of science.

Certainly, however, this is an opportunity on which it may be well for us to consider what have been the principal scientific results of the last half-century, dwelling especially on those with which this Association is more directly concerned, either as being the work of our own members, or as having been made known at our meetings. I have, moreover, especially taken those discoveries which the Royal Society has deemed worthy of a medal. It is of course impossible within the limits of a single address to do more than allude to a few of these, and that very briefly. In dealing with so large a subject I first hoped that I might take our annual volumes as a text-book. This, however, I at once found to be quite impossible. For instance, the first volume commences with a Report on Astronomy, by Sir G. Airy; I may be pardoned, I trust, for expressing my pleasure at finding that the second was one by my father, on the Tides, prepared like the preceding at the request of the Council; then comes one on Meteorology by Forbes; Radiant Heat, by Baden

Powell; Optics, by Brewster; Mineralogy, by Whewell, and so on. My best course will therefore be to take our different Sections one by one, and endeavour to bring before you a few of the principal results which have been obtained in each department.

The Biological Section is that with which I have been most intimately associated, and with which it is, perhaps, natural that I should begin.

Fifty years ago it was the general opinion that animals and plants came into existence just as we now see them. We took pleasure in their beauty; their adaptation to their habits and mode of life in many cases could not be overlooked or misunderstood. Nevertheless, the book of Nature was like some richly illuminated missal, written in an unknown tongue. The graceful forms of the letters, the beauty of the colouring, excited our wonder and admiration; but of the true meaning little was known to us; indeed we scarcely realised that there was any meaning to decipher. Now glimpses of the truth are gradually revealing themselves; we perceive that there is a reason—and in many cases we know what that reason is—for every difference in form, in size, and in colour; for every bone and every feather, almost for every hair. Moreover, each problem which is solved opens out vistas, as it were, of others perhaps even more interesting. With this important change the name of our illustrious countryman, Darwin, is intimately associated, and the year 1859 will always be memorable in science as having produced his work on 'The Origin of Species.' In the previous year he and Wallace had published short papers, in which they clearly state the theory of natural selection, at which they had simultaneously and independently arrived. We cannot wonder that Darwin's views should have at first excited great opposition. Nevertheless from the first they met with powerful support, especially, in this country, from Hooker, Huxley, and Herbert Spencer. The theory is based on four axioms :—

' 1. That no two animals or plants in nature are identical in all respects.

' 2. That the offspring tend to inherit the peculiarities of their parents.

' 3. That of those which come into existence, only a small number reach maturity.

' 4. That those, which are, on the whole, best adapted to the circumstances in which they are placed, are most likely to leave descendants.'

Darwin commenced his work by discussing the causes and extent of variability in animals, and the origin of domestic varieties; he showed the impossibility of distinguishing between varieties and species, and pointed out the wide differences which man has produced in some cases—as, for instance, in our domestic pigeons, all unquestionably descended from a common stock. He dwelt on the struggle for existence (since become a household world), which, inevitably resulting in the survival of the fittest, tends gradually to adapt any race of animals to the conditions in which it occurs.

While thus, however, showing the great importance of natural selection, he attributed to it no exclusive influence, but fully admitted that other causes—the use and disuse of organs, sexual selection, &c.—had to be taken into consideration. Passing on to the difficulties of his theory, he accounted for the absence of intermediate varieties between species, to a great extent, by the imperfection of the geological record. Here, however, I must observe that, as I have elsewhere remarked, those who rely on the absence of links between different species really argue in a vicious circle, because wherever such links do exist they regard the whole chain as a single species. The dog and jackal, for instance, are now regarded as two species, but if a series of links were discovered between them they would be united into one. Hence in this sense there never can be links between any two species, because as soon as the links are discovered the species are united. Every variable species consists, in fact, of a number of closely connected links.

But if the geological record be imperfect, it is still very instructive. The further palæontology has progressed, the more it has tended to fill up the gaps between existing groups and species: while the careful study of living forms has brought into prominence the variations dependent on food, climate, habitat, and other conditions, and shown that many species long supposed to be absolutely distinct are so closely linked together by intermediate forms that it is difficult to draw a satisfactory line between them. Thus the European and American bisons are connected by the Bison priscus of Prehistoric Europe; the grizzly bear and the brown bear, as Busk has shown, are apparently the modern representatives of the cave bear; Flower has pointed out the palæontological evidence of gradual modification of animal forms in the Artiodactyles; and we may almost say, as a general rule, that the earliest known mammalia belong to less specialised types than our existing species. They are not well-marked Carnivores, Rodents, Marsupials, &c., but rather constitute a group of generalised forms from which our present well-marked orders appear to have diverged. Among the Invertebrata, Carpenter and Williamson have proved that it is almost impossible to divide the Foraminifera into well-marked species; and, lastly, among plants, there are large genera, as, for instance, Rubus and Hieracium, with reference to the species of which no two botanists are agreed.

The principles of classification point also in the same direction, and are based more and more on the theory of descent. Biologists endeavour to arrange animals on what is called the 'natural system.' No one now places whales among fish, bats among birds, or shrews with mice, notwithstanding their external similarity; and Darwin maintained that 'community of descent was the hidden bond which naturalists had been unconsciously seeking.' How else, indeed, can we explain the fact that the framework of bones is so similar in the arm of a man, the wing of a bat, the foreleg of a horse, and the fin of a porpoise—that the neck of a giraffe and that of an elephant contain the same number of vertebræ?

Strong evidence is, moreover, afforded by embryology; by the presence of rudimentary organs and transient characters, as, for instance, the existence in the calf of certain teeth which never cut the gums, the shrivelled and useless wings of some beetles, the presence of a series of arteries in the embryos of the higher Vertebrata exactly similar to those which supply the gills in fishes, even the spots on the young blackbird, the stripes on the lion's cub; these, and innumerable other facts of the same character, appear to be incompatible with the idea that each species was specially and independently created; and to prove, on the contrary, that the embryonic stages of species show us more or less clearly the structure of their ancestors.

Darwin's views, however, are still much misunderstood. I believe there are thousands who consider that according to his theory a sheep might turn into a cow, or a zebra into a horse. No one would more confidently withstand any such hypothesis, his view being, of course, not that the one could be changed into the other, but that both are descended from a common ancestor.

No one, at any rate, will question the immense impulse which Darwin has given to the study of natural history, the number of new views he has opened up, and the additional interest which he has aroused in, and contributed to, Biology. When we were young we knew that the leopard had spots, the tiger was striped, and the lion tawny; but why this was so it did not occur to us to ask; and if we had asked no one could have answered. Now we see at a glance that the stripes of the tiger have reference to its life among jungle-grasses; the lion is sandy, like the desert; while the markings of the leopard resemble spots of sunshine glancing through the leaves. Again, Wallace in his charming essays on natural selection has shown how the same philosophy may be applied even to birds' nests—how, for instance, open nests have led to the dull colour of hen birds; the only British exception being the kingfisher, which, as we know, nests in river-banks. Lower still, among insects, Weismann has taught us that even the markings of caterpillars are full of interesting lessons; while, in other cases, specially among butterflies, Bates has made known to us the curious phenomena of mimicry.

The science of embryology may almost be said to have been created in the last half-century. Fifty years ago it was a very general opinion that animals which are unlike when mature, were dissimilar from the beginning. It is to Von Baer, the discoverer of the mammalian ovum, that we owe the great generalisation that the development of the egg is in the main a progress from the general to the special, that zoological affinity is the expression of similarity of development, and that the different great types of animal structure are the result of different modes of development—in fact, that embryology is the key to the laws of animal development.

Thus the young of existing species resemble in many cases the mature forms which flourished in ancient times. Huxley has traced up the genealogy of the horse to the Miocene Anchitherium, and his views have since been remarkably confirmed by Marsh's discovery of the Pliohippus, Protohippus, Miohippus, and Mesohippus, leading down from the Eohippus of the early tertiary strata. In the same way Boyd-Dawkins and Gaudry have called attention to the fact that just as the individual stag gradually acquires more and more complex antlers: having at first only a single prong, in the next year two points, in the following three, and so on; so the genus, as a whole, in Middle Miocene times, had two pronged horns; in the Upper Miocene, three; and that it is not till the Upper Pliocene that we find any species with the magnificent antlers of our modern deer. It seems to be now generally admitted that birds have come down to us through the Dinosaurians, and, as Huxley has shown, the profound break once supposed to exist between birds and reptiles has been bridged over by the discovery of reptilian birds and bird-like reptiles; so that, in fact, birds are modified reptiles. The remarkable genus Peripatus, so well studied by Moseley, tends to connect the annulose and articulate types.

Again, the structural resemblances between Amphioxus and the Ascidians had been pointed out by Goodsir; and Kowalevsky in 1866 showed that these were not mere analogies, but indicated a real affinity. These observations, in the words of Allen Thomson, 'have produced a change little short of revolutionary in embryological and zoological views, leading as they do to the support of the hypothesis that the Ascidian is an earlier stage in the phylogenetic history of the mammal and other vertebrates.'

The larval forms which occur in so many groups, and of which the Insects afford us the most familiar examples, are, in the words of Quatrefages, embryos, which lead an independent life. In such cases as these, external conditions act upon the larvæ as they do upon the mature form; hence we have two classes of changes, adaptational or adaptive, and developmental. These and many other facts must be taken into consideration; nevertheless naturalists are now generally agreed that embryological characters are of high value as guides in classification, and it may, I think, be regarded as well-established that, just as the contents and sequence of rocks teach us the past history of the earth, so is the gradual development of the species indicated by the structure of the embryo and its developmental changes.

When the supporters of Darwin are told that his theory is incredible, they may fairly ask why it is impossible that a species in the course of hundreds of thousands of years should have passed through changes which occupy only a few days or weeks in the life-history of each individual?

The phenomena of yolk-segmentation, first observed by Prevost and Dumas, are now known to be, in some form or other, invariably the precursors of embryonic development; while they reproduce, as the first stages in the formation of the higher animals, the main and essential features in the life-history of the lowest forms. The 'blastoderm,' as it is called, or first germ of the embryo in the egg, divides itself into two layers, corresponding, as Huxley has shown, to the two layers into which the body of the Cœlenterata may be divided. Not only so, but most embryos at an early stage of development have the form of a cup, the walls of which are formed by the two layers of the blastoderm. Kowalevsky was the first to show the prevalence of this embryonic form, and subsequently Lankester and Haeckel put forward the hypothesis that it was the embryonic repetition of an ancestral type, from which all the higher forms are descended. The cavity of the cup is supposed to be the stomach of this simple organism, and the opening of the cup the mouth. The inner layer of the wall of the cup constitutes the digestive membrane, and the outer the skin. To this form Haeckel gave the name Gastræa. It is, perhaps, doubtful whether the theory of Lankester and Haeckel can be accepted in precisely the form they propounded it; but it has had an important influence on the progress of embryology. I cannot quit the science of embryology without alluding to the very admirable work on 'Comparative Embryology' by our new general secretary, Mr. Balfour, and also the 'Elements of Embryology' which he had previously published in conjunction with Dr. M. Foster.

In 1842 Steenstrup published his celebrated work on the 'Alternation of Generations,' in which he showed that many species are represented by two perfectly distinct types or broods, differing in form, structure, and habits; that in one of them males are entirely wanting, and that the reproduction is effected by fission, or by buds, which, however, are in some cases structurally indistinguishable from eggs. Steenstrup's illustrations were mainly taken from marine or parasitic species, of very great interest, but not generally familiar, excepting to naturalists. It has since been shown that the common Cynips or Gallfly is also a case in point. It had long been known that in some genera belonging to this group, males are entirely wanting, and it has now been shown by Bassett, and more thoroughly by Adler, that some of these species are double-brooded; the two broods having been considered as distinct genera.

Thus an insect known as Neuroterus lenticularis, of which females only occur, produces the familiar oak-spangles so common on the under-sides of oak-leaves, from which emerge, not Neuroterus lenticularis, but an insect hitherto considered as a distinct species, belonging even to a different genus, Spathegaster baccarum. In Spathegaster both sexes occur; they produce the currant-like galls found on oaks, and from these galls Neuroterus is again developed. So also the King Charles oak-apples produce a species known as Teras terminalis, which descends to the ground, and makes small galls on the roots of the oak. From these emerge an insect known as Biorhiza aptera, which again gives rise to the common oak-apple.

Many butterflies, again, are dimorphic, existing under two, or even three, distinct forms—one that of the winter, the other of the summer brood or broods. Weismann has adduced strong reasons for thinking that during the glacial period these species were one-brooded only, and existed in the present winter form; that, as the climate improved, the period of warmth became sufficient to allow the development of a second brood, and led to the gradual rise of the summer form.

He and Edwards have shown that, while, by the application of cold, pupæ, which would naturally have produced the summer form, can be made to assume the winter dress; it is, on the contrary, far more difficult to change the winter into the summer colouring.

In some cases—as for instance in the very curious Leptodora crystallina (a fresh-water crustacean, inhabiting deep lakes and reservoirs, and which, as its name denotes, is almost perfectly transparent)—though the two forms are almost exactly similar in their mature state, the mode of development is very different; for, while the winter form goes through a well-marked metamorphosis, in the summer brood the development is direct.

It might seem that such enquiries as these could hardly have any practical bearing. Yet it is not improbable that they may lead to very important results. For instance, it would appear that the fluke which produces the rot in sheep, passes one phase of its existence in snails or slugs, and we are not without hopes that the researches, in which our lamented friend Prof. Rolleston was engaged at the time of his death, and which Mr. Thomas is continuing, will lead, if not to the extirpation, at any rate to the diminution, of a pest from which our farmers have so grievously suffered.

It was in the year 1839 that Schwann and Schleiden demonstrated the intimate relation in which animals and plants stand to each other, by showing the identity of the laws of development of the elementary parts in the two kingdoms of organic nature. Analogies indeed had been previously pointed out ; the presence of cellular tissue in certain parts of animals was known, but Caspar F. Wolff's brilliant memoir had been nearly forgotten; and the tendency of microscopical investigation had rather been to encourage the belief that no real similarity existed; that the cellular tissue of animals was essentially different from that of plants. This had arisen chiefly, perhaps, because fully formed tissues were compared, and it was mainly the study of the growth of cells which led to the demonstration of the general law of development for all organic elementary tissues.

As regards descriptive biology, by far the greater number of species now recorded have been named and described within the last half-century, and it is not too

INCREASE IN NUMBER OF KNOWN SPECIES. 15

much to say that not a day passes without adding new species to our lists. A comparison, for instance, of the edition of Cuvier's 'Regne Animal,' published in 1828, as compared with the present state of our knowledge, is most striking.

Dr. Günther has been good enough to make a calculation for me. The numbers, of course, are only approximate, but it appears that while the total number of animals described up to 1831 was not more than 70,000, the number now is at least 320,000.

Lastly, to show how large a field still remains for exploration, I may add that Mr. Waterhouse estimates that our Museums contain not fewer than 12,000 species of insects which have not yet been described, while our collections do not probably contain anything like one-half of those actually in existence. Further than this, the anatomy and habits even of those which have been described offer an inexhaustible field for research, and it is not going too far to say that there is not a single species which would not amply repay the devotion of a lifetime.

One remarkable feature in the modern progress of biological science has been the application of improved methods of observation and experiment; and the employment in physiological research of the exact measurements employed by the experimental physicist. Our microscopes have been greatly improved: achromatic object-glasses were introduced by Lister in 1829; the binocular arrangement by Wenham in 1856; while immersion lenses, first suggested by Amici, and since carried out under the formula of Abbe, are most valuable. The use of chemical re-agents in microscopical investigations has proved most instructive, and another very important method of investigation has been the power of obtaining very thin slices by imbedding the object to be examined in paraffin or some other soft substance. In this manner we can now obtain, say, fifty separate sections of the egg of a beetle, or the brain of a bee.

At the close of the last century, Sprengel published a most suggestive work on flowers, in which he pointed out the curious relation existing between these and insects, and showed that the latter carry the pollen from flower to flower. His observations, however, attracted little notice until Darwin called attention to the subject in 1862. It had long been known that the cowslip and primrose exist under two forms, about equally numerous, and differing from one another in the arrangements of their stamens and pistils; the one form having the stamens on the summit of the flower and the stigma half-way down; while in the other the relative positions are reversed, the stigma being at the summit of the tube and the stamens half-way down. This difference had, however, been regarded as a case of mere variability; but Darwin showed it to be a beautiful provision, the result of which is that insects fertilise each flower with pollen brought from a different plant; and he proved that flowers fertilised with pollen from the other form yield more seed than if fertilised with pollen of the same form, even if taken from a different plant.

Attention having been thus directed to the question, an astonishing variety of most beautiful contrivances have been observed and described by many botanists, especially Hooker, Axel, Delpino, Hildebrand, Bennett,

Fritz Müller, and above all Hermann Müller and Darwin himself. The general result is that to insects, and especially to bees, we owe the beauty of our gardens, the sweetness of our fields. To their beneficent, though unconscious action, flowers owe their scent and colour, their honey—nay, in many cases, even their form. Their present shape and varied arrangements, their brilliant colours, their honey, and their sweet scent are all due to the selection exercised by insects.

In these cases the relation between plants and insects is one of mutual advantage. In many species, however, plants present us with complex arrangements adapted to protect them from insects; such, for instance, are in many cases the resinous glands which render leaves unpalatable; the thickets of hairs and other precautions which prevent flowers from being robbed of their honey by ants. Again, more than a century ago, our countryman, Ellis, described an American plant, Dionæa, in which the leaves are somewhat concave, with long lateral spines, and a joint in the middle, which closes up with a jerk, like a rat-trap, the moment any unwary insect alights on them. The plant, in fact, actually captures and devours insects. This observation also remained as an isolated fact until within the last few years, when Darwin, Hooker, and others have shown that many other species have curious and very varied contrivances for supplying themselves with animal food.

As regards the progress of botany in other directions, Mr. Thiselton Dyer has been kind enough to assist me in endeavouring to place the principal facts before you. Some of the most fascinating branches of botany—

morphology, histology, and physiology scarcely existed before 1830. In the two former branches the discoveries of von Mohl are pre-eminent. He first observed cell-division in 1835, and detected the presence of starch in chlorophyll-corpuscles in 1837, while he first described protoplasm, now so familiar to us, at least by name, in 1846. In the same year Amici discovered the existence of the embryonic vesicle in the embryo sac, which develops into the embryo when fertilised by the entrance of the pollen-tube into the micropyle. The existence of sexual reproduction in the lower plants was doubtful, or at least doubted by some eminent authorities, as recently as 1853, when the actual process of fertilisation in the common bladderwrack of our shores was observed by Thuret, while the reproduction of the larger fungi was first worked out by De Bary in 1863.

As regards lichens, Schwendener proposed, in 1869, the startling theory, now however accepted by some of the highest authorities, that lichens are not autonomous organisms, but commensal associations of a fungus parasitic on an alga. With reference to the higher Cryptogams, it is hardly too much to say that the whole of our exact knowledge of their life-history has been obtained during the last half-century. Thus in the case of ferns the male organs, or antheridia, were first discovered by Nägeli in 1844, and the archegonia, or female organs, by Suminski in 1848. The early stages in the development of mosses were worked out by Valentine in 1833. Lastly, the principle of Alternation of Generations in plants was discovered by Hofmeister. This eminent naturalist also, in 1851-4, pointed out the homologies of the reproductive processes in mosses, vascular cryptogams, gymnosperms, and angiosperms.

Geographical Botany can hardly be said to have had any scientific status anterior to the publication of the 'Origin of Species.' The way had been paved, however, by A. de Candolle and the well-known essay of Edward Forbes—'On the Distribution of the Plants and Animals of the British Isles,'—by Sir J. Hooker's introductory essay to the 'Flora of New Zealand,' and by Hooker and Thomson's introductory essay to the 'Flora Indica.' One result of these researches has been to give the *coup-de-grâce* to the theory of an Atlantis. Lastly, in a lecture delivered to the Geographical Society in 1878, Thiselton Dyer himself has summed up the present state of the subject, and contributed an important addition to our knowledge of plant-distribution by showing how its main features may be explained by migration in latitude from north to south without recourse being had to a submerged southern continent for explaining the features common to South Africa, Australia, and America.

The fact that systematic and geographical botany have claimed a preponderating share of the attention of British phytologists, is no doubt in great measure due to the ever-expanding area of the British Empire, and the rich botanical treasures which we are continually receiving from India and our numerous colonies. The series of Indian and Colonial Floras, published under the direction of the authorities at Kew, and the 'Genera Plantarum' of Bentham and Hooker, are certainly an honour to our country. To similar causes we may trace the rise and rapid progress of economic botany, to which the late Sir W. Hooker so greatly contributed.

In vegetable physiology some of the most striking researches have been on the effect produced by rays of light of different refrangibility. Daubeny, Draper, and Sachs have shown that the light of the red end of the spectrum is more effective than that of the blue, so far as the decomposition of carbon dioxide (carbonic acid) is concerned.

Nothing could have appeared less likely than that researches into the theory of spontaneous generation should have led to practical improvements in medical science. Yet such has been the case. Only a few years ago Bacteria seemed mere scientific curiosities. It had long been known that an infusion—say, of hay— would, if exposed to the atmosphere, be found, after a certain time, to teem with living forms. Even those few who still believe that life would be spontaneously generated in such an infusion, will admit that these minute organisms are, if not entirely, yet mainly, derived from germs floating in our atmosphere; and if precautions are taken to exclude such germs, as in the careful experiments especially of Pasteur, Tyndall, and Roberts, everyone will grant that in ninety-nine cases out of a hundred no such development of life will take place. In 1836-7 Cagniard de la Tour and Schwann independently showed that fermentation was no mere chemical process, but was due to the presence of a microscopic plant. But, more than this, it has been gradually established that putrefaction is also the work of microscopic organisms. Thirty years, however, elapsed before these important discoveries received any practical application.

At length, however, these facts have led to most

important results in Surgery. One reason why compound fractures are so dangerous is because, the skin being broken, the air obtains access to the wound, bringing with it innumerable germs, which too often set up putrefying action. Lister first made a practical application of these observations. He set himself to find some substance capable of killing the germs without being itself too potent a caustic, and he found that dilute carbolic acid fulfilled these conditions. This discovery has enabled many operations to be performed which would previously have been almost hopeless.

The same idea seems destined to prove as useful in Medicine as in Surgery. There is great reason to suppose that many diseases, especially those of a zymotic character, have their origin in the germs of special organisms. We know that fevers run a certain definite course. The parasitic organisms are at first few, but gradually multiply at the expense of the patient, and then die out again. Indeed, it seems to be thoroughly established that many diseases are due to the excessive multiplication of microscopic organisms, and we are not without hope that means will be discovered by which, without injury to the patient, these terrible, though minute, enemies may be destroyed, and the disease thus stayed. Bacillus anthracis, for instance, is now known to be the cause of splenic fever, which is so fatal to cattle, and is also communicable to man. At Bradford, for instance, it is only too well known as the woolsorter's disease. If, however, matter containing the Bacillus be treated in a particular manner, and cattle be then inoculated with it, they are found to acquire an immunity from the fever. The interesting researches of Burdon-Sanderson, Greenfield, Koch, Pasteur, Toussaint, and others, seem to justify the hope that we may be able to modify these and other germs, and then by appropriate inoculation to protect ourselves against fever and other acute diseases.

Ferrier's researches in continuation of those of Fritsch and Hitzig have enabled us to localise the function of various parts of the brain. His results have not only proved of great importance in surgery, and in many cases led to successful operations by pointing out the exact source of the mischief, but an exact knowledge of the brain is also of the greatest importance in the treatment of nervous diseases. Echeverria has collected 165 cases of traumatic epilepsy, of which 64 per cent. were cured by removing a portion of the skull, the site for the operation and the exact locality of the injury being indicated by cerebral localisation.

The history of Anæsthetics is a most remarkable illustration how long we may be on the very verge of a most important discovery. Ether, which, as we all know, produces perfect insensibility to pain, was discovered as long ago as 1540. The anæsthetic property of nitrous oxide, now so extensively used, was observed in 1800 by Sir H. Davy, who actually experimented on himself, and had one of his teeth painlessly extracted when under its influence. He even suggests that 'as nitrous oxide gas seems capable of destroying pain, it could probably be used with advantage in surgical operations.' Nay, this property of nitrous oxide was habitually explained and illustrated in the chemical lectures given in hospitals, and yet for fifty years the gas was never used in actual operations. No one did more to promote the use of anæsthetics than Sir James Y. Simpson, who introduced chloroform, a substance which was discovered in 1831, and which for a while almost entirely superseded ether and nitrous oxide, though with improved methods of administration, the latter are now coming into favour again.

The only other reference to Physiology which time permits me to make, is the great discovery of the reflex action, as it is called, of the nervous centres. Reflex actions had been long ago observed, and it had been shown by Whytt and Hales that they were more or less independent of volition. But the general opinion was that these movements indicated some feeble power of sensation independently of the brain, and it was not till the year 1832 that the 'reflex action' of certain nervous centres was made known to us by Marshall Hall, and almost at the same period by Johannes Müller.

Few branches of science have made more rapid progress in the last half-century than that which deals with the ancient condition of Man. When our Association was founded it was generally considered that the human race suddenly appeared on the scene, about 6,000 years ago, after the disappearance of the extinct mammalia, and when Europe, both as regards physical conditions and the other animals by which it was inhabited, was pretty much in the same state as in the period covered by Greek and Roman history. Since then the persevering researches of Layard, Rawlinson, Botta and others have made known to us, not only the statues and palaces of the ancient Assyrian monarchs, but even their libraries ; the cuneiform characters have been deciphered, and we can not only see, but read, in the British Museum, the actual contemporary records, on burnt clay cylinders, of the events recorded in the historical books of the Old Testament and in the pages of Herodotus. The researches in Egypt also seem to have satisfactorily established the fact that the pyramids themselves are at least 6,000 years old, while it is obvious that the Assyrian and Egyptian monarchies cannot suddenly have attained to the wealth and power, the state of social organisation, and progress in the arts, of which we have before us, preserved by the sand of the desert from the ravages of man, such wonderful proofs.

In Europe, the writings of the earliest historians and poets indicated that, before iron came into general use, there was a time when bronze was the ordinary material of weapons, axes, and other cutting implements, and though it seemed *à priori* improbable that a compound of copper and tin should have preceded the simple metal iron, nevertheless the researches of archæologists have shown that there really was in Europe a ' Bronze Age,' which at the dawn of history was just giving way to that of ' Iron.'

The contents of ancient graves, buried in many cases so that their owner might carry some at least of his wealth with him to the world of spirits, have proved very instructive. More especially the results obtained by Nilsson in Scandinavia, by Hoare and Borlase, Bateman, Greenwell, and Pitt Rivers, in our own country, and the contents of the rich cemetery at Hallstadt, left no room for doubt as to the existence of a Bronze Age ;

but we get a completer idea of the condition of Man at this period from the Swiss lake-villages, first made known to us by Keller, and subsequently studied by Morlot, Troyon, Desor, Rütimeyer, Heer, and other Swiss archæologists. Along the shallow edges of the Swiss lakes there flourished, once upon a time, many populous villages or towns, built on platforms supported by piles, exactly as many Malayan villages are now. Under these circumstances innumerable objects were one by one dropped into the water; sometimes whole villages were burnt, and their contents submerged; and thus we have been able to recover, from the waters of oblivion in which they had rested for more than 2,000 years, not only the arms and tools of this ancient people, the bones of their animals, their pottery and ornaments, but the stuffs they wore, the grain they had stored up for future use, even fruits and cakes of bread.

But this bronze-using people were not the earliest occupants of Europe. The contents of ancient tombs give evidence of a time when metal was unknown. This also was confirmed by the evidence then unexpectedly received from the Swiss lakes. By the side of the bronze-age villages were others, not less extensive, in which, while implements of stone and bone were discovered literally by thousands, not a trace of metal was met with. The shell-mounds or refuse-heaps accumulated by the ancient fishermen along the shores of Denmark, and carefully examined by Steenstrup, Worsaae, and other Danish naturalists, fully confirmed the existence of a 'Stone Age.'

We have still much to learn, I need hardly say,

about this Stone-age people, but it is surprising how much has been made out. Evans truly observes, in his admirable work on 'Ancient Stone Implements,' 'that so far as external appliances are concerned, they are almost as fully represented as would be those of any existing savage nation by the researches of a painstaking traveller.' We have their axes, adzes, chisels, borers, scrapers, and various other tools, and we know how they made and how they used them; we have their personal ornaments and implements of war; we have their cooking utensils; we know what they ate and what they wore; lastly, we know their mode of sepulture and funeral customs. They hunted the deer and horse, the bison and urus, the bear and the wolf, but the reindeer had already retreated to the North.

No bones of the reindeer, no fragment of any of the extinct mammalia have been found in any of the Swiss lake-villages or in any of the thousands of tumuli which have been opened in our own country or in Central and Southern Europe. Yet the contents of caves and of river-gravels afford abundant evidence that there was a time when the mammoth and rhinoceros, the musk-ox and reindeer, the cave lion and hyena, the great bear and the gigantic Irish elk wandered in our woods and valleys, and the hippopotamus floated in our rivers; when England and France were united, and the Thames and the Rhine had a common estuary. This was long supposed to be before the advent of man. At length, however, the discoveries of Boucher de Perthes in the valley of the Somme, supported as they are by the researches of many continental naturalists, and in our own country of MacEnery and Godwin Austen, Prestwich and Lyell, Vivian and Pengelly, Christy, Evans, and many more, have proved that man formed a humble part of this strange assembly.

Nay, even at this early period there were at least two distinct races of men in Europe; one of them—as Boyd-Dawkins has pointed out—closely resembling the modern Esquimaux in form, in his weapons and implements, probably in his clothing, as well as in so many of the animals with which he was associated.

At this stage Man appears to have been ignorant of pottery, to have had no knowledge of agriculture, no domestic animals, except perhaps the dog. His weapons were the axe, the spear, and the javelin; I do not believe he knew the use of the bow, though he was probably acquainted with the lance. He was, of course, ignorant of metal, and his stone implements, though skilfully formed, were of quite different shapes from those of the second Stone age, and were never ground. This earlier Stone period, when man co-existed with these extinct mammalia, is known as the Palæolithic, or Early Stone Age, in opposition to the Neolithic, or Newer Stone Age.

The remains of the mammalia which co-existed with man in pre-historic times have been most carefully studied by Owen, Lartet, Rütimeyer, Falconer, Busk, Boyd-Dawkins, and others. The presence of the mammoth, the reindeer, and especially of the musk-ox, indicates a severe, not to say an arctic, climate—the existence of which, moreover, was proved by other considerations; while, on the contrary, the hippopotamus requires considerable warmth. How, then, is this association to be explained?

While the climate of the globe is, no doubt, much affected by geographical conditions, the cold of the glacial period was, I believe, mainly due to the eccentricity of the earth's orbit combined with the oblique effects of precession of the ecliptic. The result of the latter condition is a period of 21,000 years, during one half of which the northern hemisphere is warmer than the southern, while during the other 10,500 years the reverse is the case. At present we are in the former phase, and there is, we know, a vast accumulation of ice at the south pole. But when the earth's orbit is nearly circular, as it is at present, the difference between the two hemispheres is not very great; while on the contrary, as the eccentricity of the orbit increases, the contrast between them increases also. This eccentricity is continually oscillating within certain limits which Croll and subsequently Stone have calculated for the last million years. At present the eccentricity is ·016 and the mean temperature of the coldest month in London is about 40°. Such has been the state of things for nearly 100,000 years; but before that there was a period, beginning 300,000 years ago, when the eccentricity of the orbit varied from ·26 to ·57. The result of this would be greatly to increase the effect due to the obliquity of the orbit; at certain periods the climate would be much warmer than at present, while at others the number of days in winter would be twenty more, and of summer twenty less, than now, while the mean temperature of the coldest month would be lowered 20°. We thus get something like a date for the last glacial epoch, and we see that it was not simply a period of cold, but rather one of extremes, each beat of the pendulum of temperature lasting for no less than 21,000 years. This explains the fact that, as Morlot showed in 1854, the glacial deposits of Switzerland, and, as we now know, those of Scotland, are not a single uniform layer, but a succession of strata indicating very different conditions. I agree also with Croll and Geikie in thinking that these considerations explain the apparent anomaly of the co-existence in the same gravels of arctic and tropical animals; the former having lived in the cold, while the latter flourished in the hot, periods.

It is, I think, now well established that man inhabited Europe during the milder periods of the glacial epoch. Some high authorities indeed consider that we have evidence of his presence in pre-glacial and even in Miocene times, but I confess that I am not satisfied on this point. Even the more recent period carries back the record of man's existence to a distance so great as altogether to change our views of ancient history.

Nor is it only as regards the antiquity and material condition of man in pre-historic times that great progress has been made. If time permitted I should have been glad to have dwelt on the origin and development of language, of custom, and of law. On all of these the comparison of the various lower races still inhabiting so large a portion of the earth's surface, has thrown much light; while even in the most cultivated nations we find survivals, curious fancies, and lingering ideas; the fossil remains as it were of former customs and religions, embedded in our modern civilisation, like the relics of extinct animals in the crust of the earth.

In geology the formation of our Association coincided with the appearance of Lyell's 'Principles of Geology,' the first volume of which was published in 1830 and the second in 1832. At that time the received opinion was that the phenomena of Geology could only be explained by violent periodical convulsions, and a high intensity of terrestrial energy culminating in repeated catastrophes. Hutton and Playfair had indeed maintained that such causes as those now in operation, would, if only time enough were allowed, account for the geological structure of the earth; nevertheless the opposite view generally prevailed, until Lyell, with rare sagacity and great eloquence, with a wealth of illustration and most powerful reasoning, convinced geologists that the forces now in action are powerful enough, if only time be given, to produce results quite as stupendous as those which Science records.

As regards statigraphical geology, at the time of the first meeting of the British Association at York, the strata between the carboniferous limestone and the chalk had been mainly reduced to order and classified, chiefly through the labours of William Smith. But the classification of all the strata lying above the chalk and below the carboniferous limestone respectively, remained in a state of the greatest confusion. The year 1831 marks the period of the commencement of the joint labours of Sedgwick and Murchison, which resulted in the establishment of the Cambrian, Silurian, and Devonian systems. Our Pre-Cambrian strata have recently been divided by Hicks into four great groups of immense thickness, and implying a great lapse of time; but no fossils have yet been discovered in them. Lyell's classification of the Tertiary deposits; the result of the studies which he carried on with the assistance of Deshayes and others, was published in the third volume of the 'Principles of Geology' in 1833. The establishment of Lyell's divisions of Eocene, Miocene, and Pliocene, was the starting-point of a most important series of investigations by Prestwich and others of these younger deposits; as well as of the post-tertiary, quaternary, or drift beds, which are of special interest from the light they have thrown on the early history of man.

A full and admirable account of what has recently been accomplished in this department of science, especially as regards the palæozoic rocks, will be found in Etheridge's late address to the Geological Society.

The thickness of the sedimentary strata implies an enormous lapse of time, but the amount of subsequent destruction which has taken place is scarcely less surprising. Ramsay, for instance, has shown that in Wales from 9,000 to 11,000 feet of solid rock have been removed from large tracts of country. Faults or cracks there extend for miles, with the strata on one side raised in some cases as much as 10,000 feet above the same strata on the other, and yet there is not on the surface the slightest vestige of this gigantic dislocation.

The long lines of escarpment again, which stretch for miles across our country, and were long supposed to be ancient coast lines, are now ascertained, mainly through the researches of Whitaker, to be due to the differential action of aerial causes.

Before 1831 the only geological maps of this

country were William Smith's general and county maps, published between the years 1815 and 1824. In the year 1832 De la Beche made proposals to the Board of Ordnance to colour the ordnance-maps geologically, and a sum of 300*l.* was granted for the purpose. Out of this small beginning grew the important work of the Geological Survey.

The cause of slaty cleavage had long been one of the great difficulties of geology. Sedgwick suggested that it was produced by the action of crystalline or polar forces. According to this view miles and miles of country, comprising great mountain masses, were neither more nor less than parts of a gigantic crystal. Sharpe, however, called attention to the fact that shells and other fossils contained in slate rocks are compressed in a direction at right angles to the planes of cleavage, as if the rocks had been seized in the jaws of a gigantic vice. Sorby first maintained that the cleavage itself was due to pressure. He observed slate rocks containing small plates of mica, and that the effect of pressure would tend to arrange these plates with their flat surfaces perpendicular to the direction of the pressure. Tyndall has since shown that the presence of flat flakes is not necessary. He proved by experiment that pure wax could be made by pressure to split into plates of great tenuity, which he attributes mainly to the lateral sliding of the particles of the wax over each other; and thus the result of pressure on such a mass is to develop a fissile structure similar to that produced in wax on a small scale, or on a great one in the slate rocks of Cumberland or Wales.

The difficult problem of the conditions under which

granite and certain other rocks were formed was attacked by Sorby with great skill in a paper read before the Geological Society in 1858. The microscopic spaces in many minerals contain a liquid which does not entirely fill the hollow, but leaves a small vacuum ; and Sorby ingeniously pointed out that the rock must have solidified at least at a temperature high enough to expand the liquid so as to fill the cavity. Sorby's important memoir laid the foundation of microscopic petrography, which is now not only one of the most promising branches of geological research, but which has been successfully applied by Sorby himself, and by Maskelyne, to the study of meteorites.

As regards the physical character of the earth, two theories have been held : one, that of a fluid interior covered by a thin crust; the other, of a practically solid sphere. The former is now generally considered by physicists to be untenable. Though there is still much difference of opinion, the prevailing feeling on the subject has been expressed by Professor Le Conte, who says, 'the whole theory of igneous agencies—which is little less than the whole foundation of theoretic geology —must be reconstructed on the basis of a solid earth.'

In 1837 Agassiz startled the scientific world by his ' Discours sur l'ancienne extension des Glaciers,' in which, developing the observation already made by Charpentier and Venetz, that boulders had been transported to great distances, and that rocks far away from, or high above, existing glaciers, are polished and scratched by the action of ice, he boldly asserted the existence of a 'glacial period,' during which Switzerland and the North of Europe were subjected to great cold and buried under a vast sheet of ice.

The ancient poets described certain gifted mortals as privileged to descend into the interior of the earth, and have exercised their imagination in recounting the wonders there revealed. As in other cases, however, the realities of science have proved more varied and surprising than the dreams of fiction. Of the gigantic and extraordinary animals thus revealed to us, by far the greatest number have been described during the period now under review. For instance, the gigantic Cetiosaurus was described by Owen in 1338, the Dinornis of New Zealand by the same distinguished naturalist in 1839, the Mylodon in the same year, and the Archæopteryx in 1862.

In America, a large number of remarkable forms have been discovered, mainly by Marsh, Leidy, and Cope. Marsh has made known to us the Titanosaurus, of the American (Colorado) Jurassic beds, which is, perhaps, the largest land animal yet known, being a hundred feet in length, and at least thirty in height, though it seems possible that even these vast dimensions were exceeded by those of the Atlantosaurus. Nor must I omit the Hesperornis, described by Marsh in 1872, as a carnivorous, swimming ostrich, provided with teeth, which he regards as a character inherited from reptilian ancestors; the Ichthyornis, stranger still, with biconcave vertebræ, like those of fishes, and teeth set in sockets; while in the Eocene deposits of the Rocky Mountains the same indefatigable palæontologist, among other very interesting remains, has discovered three new groups of remarkable mammals, the Dinocerata, Tillodontia, and

Brontotheridæ. He has also described a number of small, but very interesting Jurassic mammalia, closely related to those found in our Stonesfield Slate and Purbeck beds, for which he has proposed a new order, 'Prototheria.' Lastly, I may mention the curiously anomalous Reptilia from South Africa, which have been made known to us by Professor Owen.

Another important result of recent palæontological research is the law of brain-growth. It is not only in the higher mammalia that we find forms with brains much larger than any existing, say, in Miocene times. The rule is almost general that—as Marsh has briefly stated it—'all tertiary mammals had small brains.' We may even carry the generalisation further. The cretaceous birds had brains one-third smaller than those of our own day, and the brain-cavities of the Dinosauria of the Jurassic period are much smaller than in any existing reptiles.

As giving, in a few words, an idea of the rapid progress in this department, I may mention that Morris's 'Catalogue of British Fossils,' published in 1843, contained 5,300 species; while that now in preparation by Mr. Etheridge enumerates 15,000.

But if these figures show how rapid our recent progress has been, they also very forcibly illustrate the imperfection of the geological record, giving us, I will not say a measure, but an idea, of the imperfection of the geological record. The number of all the described recent species is over 300,000, but certainly not half are yet on our lists, and we may safely take the total number of recent species as being not less than 700,000. But in former times there have been at the very least twelve periods, in each of which by far the greater number of species were distinct. True, the number of species was probably not so large in the earlier periods as at present; but if we make a liberal allowance for this, we shall have a total of more than 2,000,000 species, of which about 25,000 only are as yet upon record; and many of these are only represented by a few, some only by a single specimen, or even only by a fragment.

The progress of palæontology may also be marked by the extent to which the existence of groups has been, if I may so say, carried back in time. Thus I believe that in 1830 the earliest known quadrupeds were small marsupials belonging to the Stonesfield Slate; the most ancient mammal now known is Microlestes antiquus from the Keuper of Würtemberg: the oldest bird known in 1831 belonged to the period of the London Clay, the oldest now known is the Archæopteryx of the Solenhofen Slate, though it is probable that some at any rate of the footsteps on the Triassic rocks are those of birds. So again the Amphibia have been carried back from the Trias to the Coal-measures; Fish from the Old Red Sandstone to the Upper Silurian; Reptiles to the Trias; Insects from the Cretaceous to the Devonian; Mollusca and Crustacea from the Silurian to the Lower Cambrian. The rocks below the Cambrian, though of immense thickness, have afforded no relics of animal life, if we except the problematical Eozoon Canadense, so ably studied by Dawson and Carpenter. But if palæontology as yet throws no light on the original forms of life, we must remember that the simplest and the lowest organisms are so soft and perishable that they would leave 'not a wrack behind.' I will not, however, enlarge on this branch of science, because we shall have the advantage on Friday of hearing it treated with the skill of a master.

Passing to the Science of Geography, Mr. Clements Markham has recently published an excellent summary of what has been accomplished during the half-century.

As regards the Arctic regions, in the year 1830 the coast line of Arctic America was only very partially known, the region between Barrow Strait and the continent, for instance, being quite unexplored, while the eastern sides of Greenland and Spitzbergen, and the coasts of Nova Zembla, were almost unknown. Now the whole coast of Arctic America has been delineated, the remarkable archipelago to the north has been explored, and no less than seven north-west passages— none of them, however, unfortunately of any practical value—have been traced. The north-eastern passage, on the other hand, so far at least as the mouths of the great Siberian rivers, may perhaps hereafter prove of commercial importance. In the Antarctic regions, Enderby and Graham Lands were discovered in 1831–2, Balleny Islands and Sabrina Land in 1839, while the fact of the existence of the great southern continent was established in 1841 by Sir James Ross, who penetrated in 1842 to 78° 11', the southernmost point ever reached.

In Asia, to quote from Mr. Markham, ' our officers have mapped the whole of Persia and Afghanistan, surveyed Mesopotamia, and explored the Pamir steppe. Japan, Borneo, Siam, the Malay peninsula, and the greater part of China have been brought more completely to our knowledge. Eastern Turkestan has been visited, and trained native explorers have penetrated to the remotest fountains of the Oxus, and the wild plateaux of Tibet. Over the northern half of the Asiatic Continent the Russians have displayed great activity. They have traversed the wild steppes and deserts of what on old atlases was called Independent Tartary, have surveyed the courses of the Jaxartes, the Oxus, and the Amur, and have navigated the Caspian and the Sea of Aral. They have pushed their scientific investigations into the Pamir and Eastern Turkestan, until at last the British and Russian surveys have been connected.'

Again, fifty years ago the vast central regions of Africa were almost a blank upon our best maps. The rudely drawn lakes and rivers in maps of a more ancient date had become discredited. These maps did not agree among themselves, the evidence upon which they were laid down could not be found, they were in many respects highly improbable, and they seemed inconsistent with what had then been ascertained concerning the Niger and the Blue and White Niles. At the date of which I speak, the Sahara had been crossed by English travellers from the shores of the Mediterranean; but the southern desert still formed a bar to travellers from the Cape, while the accounts of traders and others who alone had entered the country from the eastern and western coasts were considered to form an insufficient basis for a map.

Since that time the successful crossing of the Kalahari desert to Lake Ngami has been the prelude to an era of African discovery. Livingstone explored

HYDROGRAPHY.

the basin of the Zambesi, and discovered vast lakes and waters which have proved to be those of the higher Congo. Burton and Speke opened the way from the West Coast, which Speke and Grant pursued into and down the Nile, and Stanley down the course of the middle and lower Congo; while the vast extension of Egyptian dominion has brought a huge slice of equatorial Africa within the limits of semi-civilisation. The western side of Africa has been attacked at many points. Alexander and Galton were among the first to make known to us its western tropical regions immediately to the north of the Cape Colony; the Ogowé has been explored; the Congo promises to become a centre of trade, and the navigable portions of the Niger, the Gambia, and the Senegal are familiarly known.

The progress of discovery in Australia has been as remarkable as that in Africa. The interior of this great continent was absolutely unknown to us fifty years ago, but is now crossed through its centre by the electric telegraph, and no inconsiderable portion of it is turned into sheep-farms. It is an interesting fact that General Sabine, so long one of our most active officers, and who is still with us, though, unfortunately his health has for some time prevented him from attending our meetings, was born on the very day that the first settler landed in Australia.

In hydrography our charts have been immensely improved. The study of rivers and of the physical geography of the sea may indeed almost be said to have come into existence as a science during the last fifty years, and in the words of Jansen, it was Maury ' who, by his wind and current charts, his trade-wind, storm, and rain charts, and last, but not least, by his work on the physical geography of the sea, gave the first great impulse to all subsequent researches.'

But the progress in our knowledge of geography is, and has been, by no means confined to the improvement of our maps, or to the discovery and description of new regions of the earth; it has extended to the causes which have led to the present configuration of the surface. To a great extent indeed this part of the subject falls rather within the scope of geology, but I may here refer, in illustration, to the distribution of lakes, the phenomena of glaciers, the formation of volcanic mountains, and the structure and distribution of coral islands.

The origin and distribution of lakes is one of the most interesting problems in physical geography. That they are not scattered at random, a glance at the map is sufficient to show. They abound in mountain districts, are comparatively rare in equatorial regions, increasing in number as we go north, so that in Scotland and the northern parts of America they are sown broadcast.

Perhaps *à priori* the first explanation of the origin of lakes which would suggest itself, would be that they were formed in hollows resulting from a disturbance of the strata, which had thrown them into a basin-shaped form. Lake-basins, however, of this character are, as a matter of fact, very rare; as a general rule lakes have not the form of basin-shaped synclinal hollows, but, on the contrary, the strike of the strata often runs right across them. My eminent predecessor, Professor Ramsay, divides lakes into three classes:—(1) Those which are due to irregular accumulations of drift, and which are generally quite shallow; (2) those which are formed by moraines; and (3) those which occupy true basins scooped by glacier-ice out of the solid rock. To the latter class belong, in his opinion, most of the great Swiss and Italian lakes. Professor Ramsay attributes their excavation to glaciers, because it is of course obvious that rivers cannot make basin-shaped hollows surrounded by rock on all sides. Now the Lake of Geneva, 1,230 feet above the sea, is 984 feet deep, the Lake of Brienz is 1,850 feet above the sea, and 2,000 feet deep, so that its bottom is really below the sea-level. The Italian lakes are even more remarkable. The Lake of Como, 700 feet above the sea, is 1,929 feet deep. Lago Maggiore, 685 feet above the sea, is no less than 2,625 feet deep. It will be observed that these lakes, like many others in mountain regions—those of Scandinavia, for instance—lie in the direct channels of the great old glaciers. If the mind is at first staggered at the magnitude of the scale, we must remember that the ice which scooped out the valley in which the Lake of Geneva now reposes, was once at least 2,700 feet thick; while the moraines were also of gigantic magnitude, that of Ivrea, for instance, being no less than 1,500 feet in height. Professor Ramsay's theory seems, therefore, to account beautifully for a large number of interesting facts.

The problem is, however, very complex; and, while admitting the force of Professor Ramsay's arguments, there are, no doubt, other causes which have exercised a considerable influence in the arrangement and configuration of lakes; for instance—as has been ably argued by our new Secretary, Mr. Bonney—irregular movements of upheaval along lines athwart the valleys.

Passing from lakes to mountains, two rival theories with reference to the structure and origin of volcanoes long struggled for supremacy.

The more general view was that the sheets of lava and scoriæ which form volcanic cones—such, for instance, as Ætna or Vesuvius—were originally nearly horizontal, and that subsequently a force operating from below, and exerting a pressure both upwards and outwards from a central axis towards all points of the compass, uplifted the whole stratified mass and made it assume a conical form, giving rise at the same time, in many cases, to a wide and deep circular opening at the top of the cone, called by the advocates of this hypothesis a 'crater of elevation.'

This theory, though, as it seems to us now, it had already received its death-blow from the admirable memoirs of Scrope, was yet that most generally adopted fifty years ago, because it was considered that compact and crystalline lavas could not have consolidated on a slope exceeding 1° or 2°. In 1858, however, Sir C. Lyell conclusively showed that in fact such lavas could consolidate at a considerable angle, even in some cases at more than 30°, and it is now generally admitted that though the beds of lava, &c., may have sustained a slight angular elevation since their deposition, still in the main, volcanic cones have acquired their form by the accumulation of lava and ashes ejected from one or more craters.

The problems presented by glaciers are of very great interest. n 1843 Agassiz and Forbes proved that the centre of a glacier, like that of a river, moves more rapidly than its sides. But how and why do glaciers move at all? Rendu, afterwards Bishop of Annecy, in 1841 endeavoured to explain the facts by supposing that glacier ice enjoys a kind of ductility. The 'viscous theory' of glaciers was also adopted, and most ably advocated by Forbes, who compared the condition of a glacier to that of the contents of a tar barrel poured into a sloping channel. We have all, however, seen long narrow fissures, a mere fraction of an inch in width, stretching far across glaciers—a condition incompatible with the ordinary idea of viscosity. The phenomenon of regelation was afterwards applied to the explanation of glacier-motion. An observation of Faraday's supplied the clue. He noticed in 1850 that when two pieces of thawing ice are placed together they unite by freezing at the place of contact. Following up this suggestion Tyndall found that if he compressed a block of ice in a mould it could be made to assume any shape he pleased. A straight prism, for instance, placed in a groove and submitted to hydraulic pressure, was bent into a transparent semicircle of ice. These experiments seem to have proved that a glacial valley is a mould through which the ice is forced, and to which it will accommodate itself, while, as Tyndall and Huxley also pointed out, the 'veined structure of ice' is produced by pressure, in the same manner as the cleavage of slate rocks.

It was in the year 1842 that Darwin published his great work on 'Coral Islands.' The fringing reefs of coral presented no special difficulty. They could be obviously accounted for by an elevation of the land, so that the coral which had originally grown under water had been raised above the sea-level. The circular or oval shape of so many reefs, however, each having a lagoon in the centre, closely surrounded by a deep ocean, and rising but a few feet above the sea-level, had long been a puzzle to the physical geographer. The favourite theory was that these were the summits of submarine volcanoes on which the coral had grown. But as the reef-making coral does not live at greater depths than about twenty-five fathoms, the immense number of these reefs formed an almost insuperable objection to this theory. The Laccadives and Maldives, for instance—meaning literally the 'lac of islands' and the 'thousand islands'—are a series of such atolls, and it was impossible to imagine so great a number of craters, all so nearly of the same altitude. Darwin showed, moreover, that so far from the ring of corals resting on a corresponding ridge of rocks, the lagoons, on the contrary, now occupy the place which was once the highest land. He pointed out that some lagoons, as for instance that of Vanikoro, contain an island in the middle; while other islands, such as Tahiti, are surrounded by a margin of smooth water separated from the ocean by a coral reef. Now if we suppose that Tahiti were to sink slowly, it would gradually approximate to the condition of Vanikoro; and if Vanikoro gradually sank, the central island would disappear, while on the contrary the growth of the coral might neutralise the subsidence of the reef, so that we should have simply an atoll with its lagoon. The same considerations explain the origin of the 'barrier reefs,' such as that which runs, for nearly one thousand miles, along the north-east coast of Australia. Thus Darwin's theory explained the form and the approximate identity of altitude of these coral islands. But it did more than this, because it showed us that there were great areas in process of subsidence, which though slow, was of great importance in physical geography.[1]

Much information has also been acquired with reference to the abysses of the ocean, especially from the voyages of the *Porcupine* and the *Challenger*. The greatest depth yet recorded is near the Ladrone Islands, where a sounding of 4,575 fathoms was obtained.

Ehrenberg long ago pointed out the similarity of the calcareous mud now accumulating in our recent seas to the chalk, and showed that the green sands of the geologist are largely made up of casts of Foraminifera. Clay, however, had been looked on, until the recent expeditions, as essentially a product of the disintegration of older rocks. Not only, however, are a large proportion of siliceous and calcareous rocks either directly or indirectly derived from material which has once formed a portion of living organisms, but Sir Wyville Thomson maintains that this is the case with some clays also. In that case the striking remark of Linnæus, that 'fossils are not the children but the parents of rocks,' will have received remarkable confirmation. I should have thought it, I confess, probable that these clays are, to a considerable extent, composed of volcanic dust.

It would appear that calcareous deposits resembling our chalk do not occur at a greater depth than 3,000

---

[1] I ought, perhaps, to mention that Darwin's views have been recently questioned on some points by Semper and Murray.

fathoms; they have not been met with in the abysses of the ocean. Here the bottom consists of exceedingly fine clay, sometimes coloured red by oxide of iron, sometimes chocolate by manganese oxide, and containing with Foraminifera occasionally large numbers of siliceous Radiolaria. These strata seem to accumulate with extreme slowness: this is inferred from the comparative abundance of whales' bones and fishes' teeth; and from the presence of minute spherical particles, supposed by Mr. Murray to be of cosmic origin—in fact, to be the dust of meteorites, which in the course of ages have fallen on the ocean. Such particles no doubt occur over the whole surface of the earth, but on land they soon oxidise, and in shallow water they are covered up by other deposits. Another interesting result of recent deep-sea explorations has been to show that the depths of the ocean are no mere barren solitudes, as was until recent years confidently believed, but, on the contrary, present us many remarkable forms of life. We have, however, as yet but thrown here and there a ray of light down into the ocean abysses:—

> Nor can so short a time sufficient be
> To fathom the vast depths of Nature's sea.

In Astronomy, the discovery in 1845 of the planet Neptune, made independently and almost simultaneously by Adams and by Le Verrier, was certainly one of the very greatest triumphs of mathematical genius. Of the minor planets four only were known in 1831, whilst the number now on the roll amounts to 220. Many astronomers believe in the existence of an intra-mercurial planet or planets, but this is still an open question. The Solar System has also been enriched by the discovery of an inner ring to Saturn, of satellites to Mars, and of additional satellites to Saturn, Uranus, and Neptune.

The most unexpected progress, however, in our astronomical knowledge during the past half-century has been due to spectrum analysis.

The dark lines in the spectrum were first seen by Wollaston, who noticed a few of them; but they were independently discovered by Fraunhofer, after whom they are justly named, and who, in 1814, mapped no fewer than 576. The first steps in 'spectrum analysis,' properly so called, were made by Sir J. Herschel, Fox Talbot, and by Wheatstone. The latter, in a paper read before this Association in 1835, showed that the spectrum emitted by the incandescent vapour of metals was formed of bright lines, and that these lines, while, as he then supposed, constant for each metal, differed for different metals. 'We have here,' he said, 'a mode of discriminating metallic bodies more readily than that of chemical examination, and which may hereafter be employed for useful purposes.' Nay, not only can bodies thus be more readily discriminated, but, as we now know, the presence of extremely minute portions can be detected, the $\frac{1}{5000000}$ of a grain being in some cases easily perceptible.

It is also easy to see that the presence of any new simple substance might be detected, and in this manner already several new elements have been discovered, as I shall mention when we come to Chemistry.

But spectrum analysis has led to even grander and more unexpected triumphs. Fraunhofer himself noticed the coincidence between the double dark line D of the solar spectrum and a double line which he observed in the spectra of ordinary flames, while Stokes pointed out to Sir W. Thomson, who taught it in his lectures, that in both cases these lines were due to the presence of sodium. To Kirchhoff and Bunsen, however, is due the independent conception and the credit of having first systematically investigated the relation which exists between Fraunhofer's lines and the bright lines in the spectra of incandescent metals. In order to get some fixed measure by which they might determine and record the lines characterising any given substance, it occurred to them that they might use for comparison the spectrum of the sun. They accordingly arranged their spectroscope so that one-half of the slit was lighted by the sun, and the other by the luminous gases they proposed to examine. It immediately struck them that the bright lines in the one corresponded with the dark lines in the other—the bright line of sodium, for instance, with the line or rather lines D·in the sun's spectrum. The conclusion was obvious. There was sodium in the sun. It must indeed have been a glorious moment when that thought flashed across them, and even by itself well worth all their labour.

But why is the bright line of a sodium flame represented by a black one in the spectrum of the sun? To Ångström is due the theory that a vapour or gas can absorb luminous rays of the same refrangibility only as those which it emits when highly heated; while Balfour Stewart independently discovered the same law with reference to radiant heat.

This is the basis of Kirchhoff's theory of the origin

of Fraunhofer's lines. In the atmosphere of the sun the vapours of various metals are present, each of which would give its characteristic lines, but within this atmospheric envelope is the still more intensely heated nucleus of the sun, which emits a brilliant continuous spectrum containing rays of all degrees of refrangibility. When the light of this intensely heated nucleus is transmitted through the surrounding atmosphere, the bright lines which would be produced by this atmosphere are seen as dark ones.

Kirchhoff and Bunsen thus proved the existence in the sun of hydrogen, sodium, magnesium, calcium, iron, nickel, chromium, manganese, titanium, and cobalt; since which Ångström, Thalen, and Lockyer have considerably increased the list.

But it is not merely the chemistry of the heavenly bodies on which light is thrown by the spectroscope; their physical structure and evolutional history are also illuminated by this wonderful instrument of research.

It used to be supposed that the sun was a dark body enveloped in a luminous atmosphere. The reverse now appears to be the truth. The body of the sun, or photosphere, is intensely brilliant; round it lies the solar atmosphere of comparatively cool gases, which cause the dark lines in the spectrum; thirdly, the chromosphere,—a sphere principally of hydrogen, jets of which are said sometimes to reach to a height of 100,000 miles or more, into the outer coating or corona, the nature of which is still very doubtful.

Formerly the red flames which represent the higher regions of the chromosphere could be seen only on the rare occasions of a total solar eclipse. Janssen and

Lockyer, by the application of the spectroscope, have enabled us to study this region of the sun at all times.

It is, moreover, obvious that the powerful engine of investigation afforded us by the spectroscope is by no means confined to the substances which form part of our system. The incandescent body can thus be examined, no matter how great its distance, so long only as the light is strong enough. That this method was theoretically applicable to the light of the stars was indeed obvious, but the practical difficulties were very great. Sirius, the brightest of all, is, in round numbers, a hundred millions of millions of miles from us; and, though as big as sixty of our suns, his light when it reaches us, after a journey of sixteen years, is at most one two-thousand-millionth part as bright. Nevertheless as long ago as 1815 Fraunhofer recognised the fixed lines in the light of four of the stars, and in 1863 Miller and Huggins in our own country, and Rutherford in America, succeeded in determining the dark lines in the spectrum of some of the brighter stars, thus showing that these beautiful and mysterious lights contain many of the material substances with which we are familiar. In Aldebaran, for instance, we may infer the presence of hydrogen, sodium, magnesium, iron, calcium, tellurium, antimony, bismuth, and mercury; some of which are not yet known to occur in the sun. As might have been expected, the composition of the stars is not uniform, and it would appear that they may be arranged in a few well-marked classes, indicating differences of temperature, or in other words, of age. Some recent photographic spectra of stars obtained by Huggins go very far to justify this view.

Thus we can make the stars teach us their own composition with light which started from its source in some cases before we were born—light older than our Association itself.

Until 1864, the true nature of the unresolved nebulæ was a matter of doubt. In that year, however, Huggins turned his spectroscope on to a nebula, and made the unexpected discovery that the spectra of some of these bodies are discontinuous—that is to say, consist of bright lines only, indicating that 'in place of an incandescent solid or liquid body we must probably regard these objects, or at least their photo-surfaces, as enormous masses of luminous gas or vapour. For it is from matter in a gaseous state only that such light as that of the nebulæ is known to be emitted.' So far as observation has yet gone, nebulæ may be divided into two classes: some giving a continuous spectrum, others one consisting of bright lines. These latter all appear to give essentially the same spectrum, consisting of a few bright lines. Two of them, in Mr. Huggins' opinion, indicate the presence of hydrogen: one of them agrees in position with a line characteristic of nitrogen.

But spectrum analysis has even more than this to tell us. The old methods of observation could determine the movements of the stars so far only as they were transverse to us; they afforded no means of measuring motion either directly towards or away from us. Now Döppler suggested in 1841 that the colors of the stars would assist us in this respect, because they would be affected by their motion to and from the earth, just as the sound of a steam-whistle is raised or lowered as it approaches or recedes from us. Everyone has observed that if a train whistles as it passes us, the sound appears to alter at the moment the engine goes by. This arises, of course, not from any change in the whistle itself, but because the number of vibrations which reach the ear in a given time are increased by the speed of the train as it approaches, and diminished as it recedes. So, like the sound, the color would be affected by such a movement; but Döppler's method was practically inapplicable, because the amount of effect on the color would be utterly insensible; and even if it were the method could not, for other reasons, be applied; indeed, as we did not know the true color of the stars, we have no datum line by which to measure.

A change of refrangibility of light, however, does occur in consequence of relative motion, and Huggins successfully applied the spectroscope to solve the problem. He took in the first place the spectroscope of Sirius, and chose a line known as F, which is due to hydrogen. Now, if Sirius was motionless, or rather if it retained a constant distance from the earth, the line F would occupy exactly the same position in the spectrum of Sirius as in that of the sun. On the contrary, if Sirius were approaching or receding from us, this line would be slightly shifted either towards the blue or red end of the spectrum. He found that the line had moved very slightly towards the red, indicating that the distance between us and Sirius is increasing at the rate of about twenty miles a second. So also Betelgeux, Rigel, Castor, and Regulus are increasing their distance, while, on the contrary, that of others, as for instance of Vega, Arcturus, and Pollux, is diminishing. The results obtained by Huggins on about twenty stars have since been confirmed and extended by Mr. Christie, now Astronomer Royal, in succession to Sir G. Airy, who has long occupied the post with so much honour to himself and advantage to science.

To examine the spectrum of a shooting star would seem even more difficult. Alexander Herschel first succeeded in doing so, and determined the presence of sodium; since which Von Konkoly has recognised the lines of magnesium, carbon, potassium, lithium, and other substances, and it appears that the shooting stars are bodies similar in character and composition to the stony masses which sometimes reach the earth as aërolites.

Some light has also been thrown upon those mysterious visitants, the comets. The researches of Prof. Newton on the periods of meteoroids led to the remarkable discovery by Schiaparelli of the identity of the orbits of some meteor-swarms with those of some comets. The similarity of orbits is too striking to be the result of chance, and shows a true cosmical relation between the bodies. Comets, in fact, are in some cases at any rate groups of meteoric stones. From the spectra of the small comets of 1866 and 1868, Huggins showed that part of their light is emitted by themselves, and reveals the presence of carbon in some form. A photographic spectrum of the comet recently visible, obtained by the same observer, is considered by him to prove that nitrogen, probably in combination with carbon, is also present.

No element has yet been found in any meteorite, which was not previously known as existing in the earth, but the phenomena which they exhibit indicate that they must have been formed under conditions very different from those which prevail on the earth's surface. I may mention, for instance, the peculiar form of crystallised silica, called by Maskelyne, Asmanite; and the whole class of meteorites, consisting of iron generally alloyed with nickel, which Daubrée terms Holosiderites. The interesting discovery, however, by Nordenskiöld, in 1870, at Ovifak, of a number of blocks of iron alloyed with nickel and cobalt, in connection with basalts containing disseminated iron, has, in the words of Judd, 'afforded a very important link, placing the terrestrial and extra-terrestrial rocks in closer relations with one another.'

We have as yet no sufficient evidence to justify a conclusion as to whether any substances exist in the heavenly bodies which do not occur in our earth, though there are many lines which cannot yet be satisfactorily referred to any terrestrial element. On the other hand, some substances which occur on our earth have not yet been detected in the sun's atmosphere.

Such discoveries as these seemed, not long ago, entirely beyond our hopes. M. Comte, indeed, in his 'Cours de Philosophie Positive,' as recently as 1842, laid it down as an axiom regarding the heavenly bodies, that ' Nous concevons la possibilité de déterminer leurs formes, leurs distances, leurs grandeurs et leurs mouvements, tandis que nous ne saurions jamais étudier par aucun moyen leur composition chimique ou leur structure minéralogique.' Yet within a few years this supposed impossibility has been actually accomplished, showing how unsafe it is to limit the possibilities of science.

It is hardly necessary to point out that, while the spectrum has taught us so much, we have still even more to learn. Why should some substances give few, and others many, lines? Why should the same substance give different lines at different temperatures? What are the relations between the lines and the physical or chemical properties?

We may certainly look for much new knowledge of the hidden actions of atoms and molecules from future researches with the spectroscope. It may even, perhaps, teach us to modify our views of the so-called simple substances. Prout, long ago, struck by the remarkable fact that nearly all atomic weights are simple multiples of the atomic weight of hydrogen, suggested that hydrogen must be the primordial substance. Brodie's researches also naturally fell in with the supposition that the so-called simple substances are in reality complex, and that their constituents occur separately in the hottest regions of the solar atmosphere. Lockyer considers that his researches lend great probability to this view. The whole subject is one of intense interest, and we may rejoice that it is occupying the attention, not only of such men as Abney, Dewar, Hartley, Liveing, Roscoe, and Schuster in our own country, but also of many foreign observers.

When geology so greatly extended our ideas of past time, the continued heat of the sun became a question of greater interest than ever. Helmholtz has shown that, while adopting the nebular hypothesis, we need not assume that the nebulous matter was originally incandescent; but that its present high temperature may be, and probably is, mainly due to gravitation between its parts. It follows that the potential energy of the sun is far from exhausted, and that with continued shrinking it will continue to give out light and heat, with little, if any, diminution for several millions of years.

Like the sand of the sea, the stars of heaven have ever been used as effective symbols of number, and the improvements in our methods of observation have added fresh force to our original impressions. We now know that our earth is but a fraction of one out of at least 75,000,000 worlds.

But this is not all. In addition to the luminous heavenly bodies, we cannot doubt that there are countless others, invisible to us from their greater distance, smaller size, or feebler light; indeed we know that there are many dark bodies which now emit no light or comparatively little. Thus in the case of Procyon, the existence of an invisible body is proved by the movement of the visible star. Again I may refer to the curious phenomena presented by Algol, a bright star in the head of Medusa. The star shines without change for two days and thirteen hours; then, in three hours and a half, dwindles from a star of the second to one of the fourth magnitude; and then, in another three and a half hours, reassumes its original brilliancy. These changes seem certainly to indicate the presence of an opaque body, which intercepts at regular intervals a part of the light emitted by Algol.

Thus the floor of heaven is not only 'thick inlaid with patines of bright gold,' but studded also with extinct stars, once probably as brilliant as our own sun, but now dead and cold, as Helmholtz tells us that our sun itself will be, some seventeen millions of years hence.

The connection of Astronomy with the history of our planet has been a subject of speculation and research during a great part of the half-century of our existence. Sir Charles Lyell devoted some of the opening chapters of his great work to the subject. Haughton has brought his very original powers to bear on the subject of secular changes in climate, and Croll's contributions to the same subject are of great interest. Last, but not least, I must not omit to make mention of the series of massive memoirs (I am happy to say not yet nearly terminated) by George Darwin on tidal friction, and the influence of tidal action on the evolution of the solar system.

I may perhaps just mention, as regards telescopes, that the largest reflector in 1830 was Sir W. Herschel's of 4 ft., the largest at present being Lord Rosse's of 6 ft.; as regards refractors the largest then had a diameter of $11\frac{1}{4}$ in., while your fellow-townsman Cooke carried the size to 25 in., and Mr. Grubb, of Dublin, has just successfully completed one of 27 in. for the Observatory of Vienna. It is remarkable that the two largest telescopes in the world should both be Irish.

The general result of astronomical researches has been thus eloquently summed up by Proctor :—' The sidereal system is altogether more complicated and more varied in structure than has hitherto been supposed ; in the same region of the stellar depths co-exist stars of many orders of real magnitude ; all orders of nebulæ, gaseous or stellar, planetary, ring-formed, elliptical, and spiral, exist within the limits of the galaxy; and lastly, the whole system is alive with movements, the laws of which may one day be recognised, though at present they appear too complex to be understood.'

We can, I think, scarcely claim the establishment of the undulatory theory of light as falling within the last fifty years; for though Brewster, in his 'Report on Optics,' published in our first volume, treats the question as open, and expresses himself still unconvinced, he was, I believe, almost alone in his preference for the emission theory. The phenomena of interference, in fact, left hardly any—if any—room for doubt, and the subject was finally set at rest by Foucault's celebrated experiments in 1850. According to the undulatory theory the velocity of light ought to be greater in air than in water, while if the emission theory were correct the reverse would be the case. The velocity of light—186,000 miles in a second—is, however, so great that, to determine its rate in air, as compared with that in water, might seem almost hopeless. The velocity in air was, nevertheless, determined by Fizeau in 1849, by means of a rapidly revolving wheel. In the following year Foucault, by means of a revolving mirror, demonstrated that the velocity of light is greater in air than in water—thus completing the evidence in favour of the undulatory theory of light.

The idea is now gaining ground, that, as maintained by Clerk-Maxwell, light itself is an electromagnetic disturbance, the luminiferous ether being the vehicle of both light and electricity.

Wünsch, as long ago as 1792, had clearly shown

that the three primary colors were red, green, and violet; but his results attracted little notice, and the general view used to be that there were seven principal colors—red, orange, yellow, green, blue, indigo, and violet; four of which—namely orange, green, indigo, and violet—were considered to arise from mixtures of the other three. Red, yellow, and blue were therefore called the primary colors, and it was supposed that in order to produce white light these three colors must always be present.

Helmholtz, however, again showed, in 1852, that a color to our unaided eyes identical with white, was produced by combining yellow with indigo. At that time yellow was considered to be a simple color, and this, therefore, was regarded as an exception to the general rule, that a combination of three simple colors is required to produce white. Again, it was, and indeed still is, the general impression that a combination of blue and yellow makes green. This, however, is entirely a mistake. Of course we all know that yellow paint and blue paint make green paint: but this results from absorption of light by the semi-transparent solid particles of the pigments, and is not a mere mixture of the colors proceeding unaltered from the yellow and the blue particles; moreover, as can easily be shown by two sheets of colored paper and a piece of window glass, blue and yellow light, when combined, do not give a trace of green, but if pure would produce the effect of white. Green, therefore, is after all not produced by a mixture of blue and yellow. On the other hand, Clerk-Maxwell proved in 1860 that yellow could be produced by a mixture of red and green, which put an end to the pretension of yellow to be considered a primary element of color. From these and other considerations it would seem, therefore, that the three primary colors—if such an expression be retained—are red, green, and violet.

The existence of rays beyond the violet, though almost invisible to our eyes, had long been demonstrated by their chemical action. Stokes, however, showed in 1852 that their existence might be proved in another manner, for that there are certain substances which, when excited by them, emit light visible to our eyes. To this phenomenon he gave the name of fluorescence. At the other end of the spectrum Abney has recently succeeded in photographing a large number of lines in the infra-red portion, the existence of which was first proved by Sir William Herschel.

From the rarity, and in many cases the entire absence, of reference to blue, in ancient literature, Geiger—adopting and extending a suggestion first thrown out by Mr. Gladstone—has maintained that, even as recently as the time of Homer, our ancestors were blue-blind. Though for my part I am unable to adopt this view, it is certainly very remarkable that neither the Rigveda, which consists almost entirely of hymns to heaven, nor the Zendavesta, the Bible of the Parsees or fire-worshippers, nor the earlier books of the Old Testament, nor the Homeric poems, ever allude to the sky as blue.

On the other hand, from the dawn of poetry, the splendours of the morning and evening skies have excited the admiration of mankind. As Ruskin says, in language almost as brilliant as the sky itself, the whole heaven, 'from the zenith to the horizon, becomes one molten, mantling sea of color and fire; every black bar turns into massy gold, every ripple and wave into unsullied shadowless crimson, and purple, and scarlet, and colors for which there are no words in language, and no ideas in the mind—things which can only be conceived while they are visible; the intense hollow blue of the upper sky melting through it all, showing here deep, and pure, and lightness; there, modulated by the filmy, formless body of the transparent vapour, till it is lost imperceptibly in its crimson and gold.'

But what is the explanation of these gorgeous colors? why is the sky blue? and why are the sunrise and sunset crimson and gold? It may be said that the air is blue, but if so how can the clouds assume their varied tints? Brücke showed that very minute particles suspended in water are blue by reflected light. Tyndall has taught us that the blue of the sky is due to the reflection of the blue rays by the minute particles floating in the atmosphere. Now if from the white light of the sun the blue rays are thus selected, those which are transmitted will be yellow, orange, and red. Where the distance is short the transmitted light will appear yellowish. But as the sun sinks towards the horizon the atmospheric distance increases, and consequently the number of the scattering particles. They weaken in succession the violet, the indigo, the blue, and even disturb the proportions of green. The transmitted light under such circumstances must pass from yellow through orange to red, and thus, while we at noon are admiring the deep blue of the sky, the same rays, robbed of their blue, are elsewhere lighting up the evening sky with all the glories of sunset.

Another remarkable triumph of the last half-century has been the discovery of photography. At the commencement of the century Wedgwood and Davy observed the effect produced by throwing the images of objects on paper or leather prepared with nitrate of silver, but no means were known by which such images could be fixed. This was first effected by Niepce, but his processes were open to objections, which prevented them from coming into general use, and it was not till 1839 that Daguerre invented the process which was justly named after him. Very soon a further improvement was effected by our countryman Talbot. He not only fixed his 'Talbotypes' on paper—in itself a great convenience—but, by obtaining a negative, rendered it possible to take off any number of positive, or natural, copies from one original picture. This process is the foundation of all the methods now in use; perhaps the greatest improvements having been the use of glass plates, first proposed by Sir John Herschel; of collodion, suggested by Le Grey, and practically used by Archer; and, more lately, of gelatine, the foundation of the sensitive film now growing into general use in the ordinary dry-plate process. Not only have a great variety of other beautiful processes been invented, but the delicacy of the sensitive film has been immensely increased, with the advantage, among others, of diminishing greatly the time necessary for obtaining a picture, so that even an express train going at full speed can now be taken. Indeed, with full sunlight $\frac{1}{600}$ of a second is enough, and in photographing the sun itself $\frac{1}{60000}$ of a second is sufficient.

We owe to Wheatstone the conception that the idea

of solidity is derived from the combination of two pictures of the same object in slightly different perspective. This he proved in 1833 by drawing two outlines of some geometrical figure or other simple object, as they would appear to either eye respectively, and then placing them so that they might be seen, one by each eye. The 'stereoscope,' thus produced, has been greatly popularised by photography.

For 2,000 years the art of lighting had made little if any progress. Until the close of the last century, for instance, our lighthouses contained mere fires of wood or coal, though the construction had vastly improved. The Eddystone lighthouse, for instance, was built by Smeaton in 1759 ; but for forty years its light consisted of a row of tallow candles stuck in a hoop. The Argand lamp was the first great improvement, followed by gas, and in 1863 by the electric light.

Just as light was long supposed to be due to the emission of material particles, so heat was regarded as a material, though ethereal, substance, which was added to bodies when their temperature was raised.

Davy's celebrated experiment of melting two pieces of ice by rubbing them against one another in the exhausted receiver of an air-pump had convinced him that the cause of heat was the motion of the invisible particles of bodies, as had been long before suggested by Newton, Boyle, and Hooke. Rumford and Young also advocated the same view. Nevertheless, the general opinion, even until the middle of the present century, was that heat was due to the presence of a subtle fluid known as 'caloric,' a theory which is now entirely abandoned.

Melloni, by the use of the electric pile, vastly increased our knowledge of the phenomena of radiant heat. His researches were confined to the solid and liquid forms of matter. Tyndall studied the gases in this respect, showing that differences greater than those established by Melloni existed between gases and vapours, both as regards the absorption and radiation of heat. He proved, moreover, that the aqueous vapour of our atmosphere, by checking terrestrial radiation, augments the earth's temperature, and he considers that the existence of tropical vegetation—the remains of which now constitute our coal-beds—may have been due to the heat retained by the vapours which at that period were diffused in the earth's atmosphere. Indeed, but for the vapour in our atmosphere, a single night would suffice to destroy the whole vegetation of the temperate regions.

Inspired by a contemplation of Graham Bell's ingenious experiments with intermittent beams on solid bodies, Tyndall took a new and original departure; and regarding the sounds as due to changes of temperature he concluded that the same method would prove applicable to gases. He thus found himself in possession of a new and independent method of procedure. It need perhaps be hardly added that, when submitted to this new test, his former conclusions on the interaction of heat and gaseous matter stood their ground.

The determination of the mechanical equivalent of heat is mainly due to the researches of Mayer and Joule. Mayer, in 1842, pointed out the mechanical equivalent of heat as a fundamental datum to be determined by experiment. Taking the heat produced by the condensation of air as the equivalent of the work done in compressing the air, he obtained a numerical value of the mechanical equivalent of heat. There was, however, in these experiments, one weak point. The matter operated on did not go through a cycle of changes. He assumed that the production of heat was the only effect of the work done in compressing the air. Joule had the merit of being the first to meet this possible source of error. He ascertained that a weight of 1 lb. would have to fall 772 feet in order to raise the temperature of 1 lb. of water by 1° Fahr. Hirn subsequently attacked the problem from the other side, and showed that if all the heat passing through a steam engine was turned into work, for every degree Fahr. added to the temperature of a pound of water, enough work could be done to raise a weight of 1 lb. to a height of 772 feet. The general result is that, though we cannot create energy we may help ourselves to any extent from the great storehouse of nature. Wind and water, the coal-bed and the forest, afford man an inexhaustible supply of available energy. It used to be considered that there was an absolute break between the different states of matter. The continuity of the gaseous, liquid, and solid conditions was first demonstrated by Andrews in 1862.

Oxygen and nitrogen have been liquefied independently, and at the same time, by Cailletet and Raoul Pictet. Cailletet also succeeded in liquefying air, and soon afterwards hydrogen was liquefied by Pictet under a pressure of 650 atmospheres, and a cold of 170° Cent. below zero. It even became partly solidified, and he assures us that it fell on the floor with 'the shrill noise of metallic hail.' Thus then it was shown experimentally that there are no such things as absolutely permanent gases.

The kinetic theory of gases, now generally accepted, refers the elasticity of gases to a motion of translation of their molecules, and we are assured that in the case of hydrogen at a temperature of 60° Fahr. they move at an average rate of 6,225 feet in a second; while as regards their size, Loschmidt, who has since been confirmed by Stoney and Sir W. Thomson, calculates that each is at most $\frac{1}{50000000}$ of an inch in diameter.

We cannot, it would seem at present, hope for any increase of our knowledge of atoms by any improvement in the microscope. With our present instruments we can perceive lines ruled on glass of $\frac{1}{90000}$th of an inch apart. But, owing to the properties of light itself, the fringes due to interference begin to produce confusion at distances of $\frac{1}{74000}$, and in the brightest part of the spectrum at little more than $\frac{1}{90000}$th they would make the obscurity more or less complete. If indeed we could use the blue rays by themselves, their waves being much shorter, the limit of possible visibility might be extended to $\frac{1}{120000}$; and as Helmholtz has suggested, this perhaps accounts for Stinde having actually been able to abtain a photographic image of lines only $\frac{1}{100000}$th of an inch apart. It would seem then that, owing to the physical characters of light, we can, as Sorby has pointed out, scarcely hope for any great improvement so far as the mere visibility of structure is concerned, though in other respects no doubt much may be hoped for. At the same time, Dallinger and Royston Pigott have shown that, so far as the mere

THE SIZE OF MOLECULES. METEOROLOGY. 67

presence of simple objects is concerned, bodies of even smaller dimensions can be perceived.

According to the views of Helmholtz, the smallest particle that could be distinctly defined, when associated with others, is about $\frac{1}{80000}$th of an inch. Sorby estimates that a particle of albumen of this size contains 125,000,000 of molecules. In the case of such a simple compound as water the number would be no less than 8,000,000,000. Even, then, if we could construct microscopes far more powerful than any we now possess, they would not enable us to obtain by direct vision any idea of the ultimate molecules of matter. Sorby calculates that the smallest sphere of organic matter which could be clearly defined with our most powerful microscopes would contain many millions of molecules of albumen and water, and it follows that there may be an almost infinite number of structural characters in organic tissues, which we can at present foresee no mode of examining.

The science of Meteorology has made great progress; the weather, which was formerly treated as a local phenomenon, being now shown to form part of a vast system of mutually dependent cyclonic and anti-cyclonic movements. The storm-signals issued at our ports are very valuable to sailors, while the small weather-maps, for which we are mainly indebted to Francis Galton, and the forecasts, which anyone can obtain on application either personally or by telegraph at the Meteorological Office, are also of increasing utility.

Electricity in the year 1831 may be considered to have just been ripe for its adaptation to practical purposes; it was but a few years previously, in 1819, that Oersted had discovered the deflective action of the current on the magnetic needle, that Ampère had laid the foundation of electro-dynamics, that Schweizzer had devised the electric coil or multiplier, and that Sturgeon had constructed the first electro-magnet. It was in 1831 that Faraday, the prince of pure experimentalists, announced his discoveries of voltaic induction and magneto-electricity, which with the other three discoveries constitute the principles of nearly all the telegraph instruments now in use; and in 1834 our knowledge of the nature of the electric current had been much advanced by the interesting experiment of Sir Charles Wheatstone, proving the velocity of the current in a metallic conductor to approach that of the wave of light.

Practical applications of these discoveries were not long in coming to the fore, and the first telegraph line on the Great Western Railway from Paddington to West Drayton was set up in 1838. In America, Morse is said to have commenced the development of his recording instrument between the years 1832 and 1837, while Steinheil, in Germany, during the same period was engaged upon his somewhat super-refined ink-recorder, using for the first time the earth for completing the return circuit; whereas in this country Cooke and Wheatstone, by adopting the more simple device of the double-needle instrument, were the first to make the electric telegraph a practical institution. Contemporaneously with, or immediately succeeding these pioneers, we find in this country Alexander Bain, Breguet in France, Schilling in Russia, and Werner

Siemens in Germany, the last having first, in 1847, among others, made use of gutta-percha as an insulating medium for electric conductors, and thus cleared the way for subterranean and submarine telegraphy.

Four years later, in 1851, submarine telegraphy became an accomplished fact through the successful establishment of telegraphic communication between Dover and Calais. Submarine lines followed in rapid succession, crossing the English Channel and the German Ocean, threading their way through the Mediterranean, Black and Red Seas, until in 1866, after two abortive attempts, telegraphic communication was successfully established between the Old and New Worlds, beneath the Atlantic Ocean.

In connection with this great enterprise and with many investigations and suggestions of a highly scientific and important character, the name of Sir William Thomson will ever be remembered. The ingenuity displayed in perfecting the means of transmitting intelligence through metallic conductors, with the utmost despatch and certainty as regards the record obtained, between two points hundreds and even thousands of miles apart is truly surprising. The instruments devised by Morse, Siemens, and Hughes have also proved most useful.

Duplex and quadruplex telegraphy, one of the most striking achievements of modern telegraphy, the result of the labours of several inventors, should not be passed over in silence. It not only serves for the simultaneous communication of telegraphic intelligence in both directions, but renders it possible for four instruments to be worked irrespectively of one another, through one and the same wire connecting two distant places.

Another more recent and perhaps still more wonderful achievement in modern telegraphy is the invention of the telephone and microphone, by means of which the human voice is transmitted through the electric conductor, by mechanism that imposes through its extreme simplicity. In this connection the names of Reiss, Graham Bell, Edison, and Hughes are those chiefly deserving to be recorded.

Whilst electricity has thus furnished us with the means of flashing our thoughts by record or by voice from place to place, its use is now gradually extending for the achievement of such quantitative effects as the production of light, the transmission of mechanical power, and the precipitation of metals. The principle involved in the magneto-electric and dynamo-electric machines, by which these effects are accomplished, may be traced to Faraday's discovery in 1831 of the induced current, but their realisation to the labours of Holmes, Siemens, Pacinotti, Gramme, and others. In the electric light, gas-lighting has found a formidable competitor, which appears destined to take its place in public illumination, and in lighting large halls, works, &c., for which purposes it combines brilliancy and freedom from obnoxious products of combustion, with comparative cheapness. The electric light seems also to threaten, when sub-divided in the manner recently devised by Edison, Swan, and others, to make inroads into our dwelling-houses.

By the electric transmission of power, we may hope some day to utilise at a distance such natural sources of energy as the Falls of Niagara, and to work our cranes, lifts, and machinery of every description by means of sources of power arranged at convenient centres. To these applications the Brothers Siemens have more recently added the propulsion of trains by currents passing through the rails, the fusion in considerable quantities of highly refractory substances, and the use of electric centres of light in horticulture as proposed by Werner and William Siemens. By an essential improvement by Faure of the Planté Secondary Battery, the problem of storing electrical energy appears to have received a practical solution, the real importance of which is clearly proved by Sir W. Thomson's recent investigation of the subject.

It would be difficult to assign the limits to which this development of electrical energy may not be rendered serviceable for the purposes of man.

As regards mathematics I have felt that it would be impossible for me, even with the kindest help, to write anything myself. Mr. Spottiswoode, however, has been so good as to supply me with the following memorandum.

> In a complete survey of the progress of science during the half-century which has intervened between our first and our present meeting, the part played by mathematics would form no insignificant feature. To those indeed who are outside its enchanted circle it is difficult to realise the intense intellectual energy which actuates its devotees, or the wide expanse over which that energy ranges. Some measure, however, of its progress may perhaps be formed by considering, in one or two cases, from what simple principles some of the great recent developments have taken their origin.
> Consider, for instance, what is known as the principle of signs. In geometry we are concerned with quantities such as lines and angles;

and in the old systems a proposition was proved with reference to a particular figure. This figure might, it is true, be drawn in any manner within certain ranges of limitation: but if the limits were exceeded, a new proof, and often a new enunciation, became necessary. Gradually, however, it came to be perceived (*e.g.* by Carnot, in his 'Géometrie de Position,') that some propositions were true even when the quantities were reversed in direction. Hence followed a recognition of the principle (of signs) that every line should be regarded as a directed line, and every angle as measured in a definite direction. By means of this simple consideration, geometry has acquired a power similar to that of algebra, viz. of changing the signs of the quantities and transposing their positions, so as at once, and without fresh demonstration, to give rise to new propositions.

To take another instance. The properties of triangles, as established by Euclid, have always been considered as legitimate elements of proof; so that, when in any figure two triangles occur, their relations may be used as steps in a demonstration. But, within the period of which I am speaking, other general geometrical relations, *e.g.* those of a pencil of rays, or of their intersection with a straight line, have been recognised as serving a similar purpose. With what extensive results this generalisation has been attended, the Géometrie Supérieure of the late M. Chasles, and all the superstructure built on Anharmonic Ratio as a foundation, will be sufficient evidence.

Once more, the algebraical expression for a line or a plane involves two sets of quantities, the one relating to the position of any point in the line or plane, and the other relating to the position of the line or plane in space. The former set alone were originally considered variable, the latter constant. But as soon as it was seen that either set might at pleasure be regarded as variable, there was opened out to mathematicians the whole field of duality within geometry proper, and the theory of correlative figures which is destined to occupy a prominent position in the domain of mathematics.

Not unconnected with this is the marvellous extension which the transformation of geometrical figures has received very largely from Cremona and the Italian School, and which in the hands of our countrymen Hirst and the late Professor Clifford, has already brought forth such abundant fruit. In this, it may be added, there lay—dormant, it is true, and long unnoticed—the principle whereby circular may be converted into rectilinear motion, and *vice versâ*—a problem which until the time of Peaucillier seemed so far from solution, that one of the greatest mathematicians of the day thought that he had proved its entire impossibility. In the hands of Sylvester, of Kempe, and others, this principle has been developed into a general theory of link-work, on which the last word has not yet been said.

If time permitted, I might point out how the study of particular geometric figures, such as curves and surfaces, has been in many instances replaced by that of systems of figures infinite in number, and indeed of various degrees of infinitude. Such, for instance, are Plücker's complexes and congruencies. I might describe also how Riemann taught us that surfaces need not present simple extension without thickness; but that, without losing their essential geometric character, they may consist of manifold sheets; and thus our conception of space, and our power of interpreting otherwise perplexing algebraical expressions, become immensely enlarged.

Other generalisations might be mentioned, such as the principle of continuity, the use of imaginary quantities, the extension of the number of the dimensions of space, the recognition of systems in which the axioms of Euclid have no place. But as these were discussed in a recent address, I need not now do more than remind you that the germs of the great calculus of Quaternions were first announced by their author, the late Sir W. R. Hamilton, at one of our meetings.

Passing from geometry proper to the other great branch of mathematical machinery, viz. algebra, it is not too much to say that within the period now in review there has grown up a modern algebra which to our founders would have appeared like a confused dream, and whose very language and terminology would be as an unknown tongue.

Into this subject I do not propose to lead you far. But, as the progress which has been made in this direction is certainly not less than that made in geometry, I will ask your attention to one or two points which stand notably prominent.

In algebra we use ordinary equations involving one unknown quantity; in the application of algebra to geometry we meet with equations, representing curves or surfaces, and involving two, or three, unknown quantities respectively; in the theory of probabilities, and in other branches of research, we employ still more general expressions. Now the modern algebra, originating with Cayley and Sylvester, regards all these diverse expressions as belonging to one and the same family, and comprises them all under the same general term 'quantics.' Studied from this point of view, they all alike give rise to a class of derivative forms, previously unnoticed, but now known as invariants, covariants, canonical forms, etc. By means of these, mathematicians have arrived not only at many properties of the quantics themselves, but also, at their application to physical problems. It would be a long and perhaps invidious task to enumerate the many workers in this fertile field of research, especially in the schools of Germany and of Italy; but it is perhaps the less necessary to do so, because Sylvester, aided by a young and vigorous staff at Baltimore, is welding many of these results into a homogeneous mass in the classical memoirs which are appearing from time to time in the American 'Journal of Mathematics.'

In order to remove any impression that these extensions of algebra are merely barren speculations of ingenious intellects, I may add that many of these derivative forms, at least in their elementary stages, have already found their way into the text-books of mathematics; and one class in particular, known by the name of determinants, is now introduced as a recognised method of algebra, greatly to the convenience of all those who become master of its use.

In the extension of mathematics it has happened more than once that laws have been established so simple in form, and so obvious in their necessity, as scarcely to require proof. And yet their application is often of the highest importance in checking conclusions which have been drawn from other considerations, as well as in leading to conclusions which, without their aid, might have been difficult of attainment. The same thing has occurred also in physics; and notably in the recognition of what has been termed the 'Law of the Conservation of Energy.'

Energy has been defined to be 'The capacity, or power, of any body, or system of bodies, when in a given condition, to do a measurable quantity of work.' Such work may either change the condition of the bodies in question, or it may affect other bodies; but in either case energy is expended by the agent upon the recipient in performance of the work. The law then states that the total amount of energy in the agents and recipients taken together remains unaltered by the changes in question.

Now the principle on which the law depends is this:—'that every kind of change among the bodies may be expressed numerically in one standard unit of change, viz., work done, in such wise that the result of the passage of any system from one condition to another may be calculated by mere additions and substractions, even when we do not know how the change came about. This being so, all work done by a system may be expressed as a diminution of energy of that system, and all work done upon a system as an accession of energy. Consequently, the energy lost by one system in performance of work will be gained by another in having work done upon it, and the total energy, as between the two systems, will remain unchanged.

There are two cases, or conditions, of energy which, although substantially the same, are for convenience regarded separately. These may be illustrated by the following example. Work may be done upon a body, and energy communicated to it, by setting it in motion, *e.g.* by lifting it against gravity. Suppose this to be done by a spring and detent; and suppose further the body, on reaching its highest point, to be caught so as to rest at that level on a support. Then, whether we consider the body at the moment of starting, or when resting on the support, it has equally received an accession of energy from the spring, and is therefore equally capable of communicating energy to a third body. But in the one case this is due to the motion which it has acquired, and in the other to the position at which it rests, and to its capability of falling again when the support is removed. Energy in the first of these states is called 'Energy of Motion,' or 'Kinetic Energy,' and that in the second state, 'Energy of Position,' or 'Potential Energy.' In the case supposed, at the moment of starting, the whole of the energy is kinetic; as the body rises, the energy becomes partly potential and partly kinetic; and when it reaches the highest point the energy has become wholly potential. If the body be again dropped, the process is reversed.

The history of a discovery, or invention, so simple at first sight, is often found to be more complicated the more thoroughly it is examined. That which at first seems to have been due to a single mind proves to have been the result of the successive actions of many minds. Attempts more or less successful in the same direction are frequently traced out; and even unsuccessful efforts may not have been without influence on minds turned towards the same object. Lastly also, germs of thought, originally not fully understood, sometimes prove in the end to have been the first stages of growth towards ultimate fruit. The history of the law of the conservation of energy forms no exception to this order of events. There are those who discern even in the writings of Newton expressions which show that he was in possession of some ideas which, if followed out in a direct line of thought, would lead to those now entertained on the subjects of energy and of work. But however this may be, and whosoever might be reckoned among the earlier contributors to the general subject of energy, and to the establishment of its laws, it is certain that within the period of which

I am now speaking, the names of Séguin, Clausius, Helmholtz, Mayer, and Colding on the Continent, and those of Grove, Joule, Rankine, and Thomson in this country, will always be associated with this great work.

I must not, however, quit this subject without a passing notice of a conclusion to which Sir William Thomson has come, and in which he is followed by others who have pursued the transformation of energy to some of its ultimate consequences. The nature of this will perhaps be most easily apprehended by reference to a single instance. In a steam engine, or other engine, in which the motive power depends upon heat, it is well known that the source of power lies not in the general temperature of the whole, but merely on the difference of temperature between that of the boiler and that of the condenser. And the effect of the condenser is to reduce the steam issuing from the boiler to the same temperature as that of the condenser. When this is once done, no more work can be got out of the engine, unless fresh heat be supplied from an outside source to the boiler. The heat originally communicated to the boiler has become uniformly diffused, and the energy due to that difference is said to have been dissipated. The energy remains in a potential condition as regards other bodies; but as regards the engine, it is of no further use. Now suppose that we regard the entire material universe as a gigantic engine, and that after long use we have exhausted all the fuel (in its most general sense) in the world; then all the energy available will have become dissipated, and we shall have arrived at a condition of things from which there is no apparent escape. This is what is called the 'Dissipation of Energy.'

Prof. Frankland has been so good as to draw up for me the following account of the progress of Chemistry during the last half-century.

Most of the elements had been discovered before 1830, the majority of the rarer elements since the beginning of the century. In addition to these the following five have been discovered, three of them by Mosander, viz. :—lanthanum in 1839, didymium in 1842, and ebrium in 1843. Ruthenium was discovered by Claus in 1843, and niobium by Rose in 1844. Spectrum analysis has added five to the list, viz. :—Cæsium and rubidium, which were discovered by Bunsen and Kirchhoff in 1860; thallium, by Crookes in 1861; indium by Reich and Richter in 1863 ; and gallium, by Lecoq de Boisbaudran in 1875.

As regards theoretical views, the atomic theory, the foundation of scientific chemistry, had been propounded by Dalton (1804-1808). The three laws which have been chiefly instrumental in establishing the true atomic weights of the elements—the law of Avogadro (1811), that equal volumes of gases under the same conditions of temperature and pressure contain equal numbers of molecules; the law of Dulong and Petit (1819), that the capacities for heat of the atoms of the various elements are equal; and Mitscherlich's law of isomorphism (1819), according to which equal numbers of atoms of elements belonging to the same class may replace each other in a compound without altering the crystalline form of the latter, had been enunciated in quick succession; but the true application of these three laws, though in every case distinctly stated by the discoverers, failed to be generally made, and it was not till the rectification of the atomic weights by Cannizzaro, in 1858, that these important discoveries bore fruit.

In organic chemistry the views most generally held about the year 1830 were expressed in the radical theory of Berzelius. This theory, which was first stated in its electro-chemical and dualistic form by its author in 1817, received a further development at his hands in 1834 after the discovery of the benzoyl-radical by Liebig and Wöhler. In the same year (1834), however, a discovery was made by Dumas, which was destined profoundly to modify the electro-chemical portion of the theory, and even to overthrow the form of it put forth by Berzelius. Dumas showed that an electro-negative element, such as chlorine, might replace, atom for atom, an electro-positive element like hydrogen, in some cases without much alteration in the character of the compound. This law of substitution has formed a necessary portion of every chemical theory which has been proposed since its discovery, and its importance has increased with the progress of the science. It would take too long to enumerate all the theoretical views which have prevailed at various times during the past fifty years; but the theory which along with the radical theory has exercised most influence on the development of the views now held, is the theory of types, first stated by Dumas (1839) and developed in a different form and amalgamated with the radical theory by Gerhardt and Williamson (1848-1852). It is, however, the less necessary to refer in detail to these views, seeing that in the now prevailing theory of atomicity we possess a generalisation which, while greatly extending the scope of chemical science in its power of classifying known and predicting unknown facts, includes all that was valuable in the generalisations which preceded it. The study of the behaviour of organo-metallic compounds in chemical reactions led to the conclusion that various metallic elements possess a definite capacity of saturation with regard to the number of atoms of other elements with which they can combine, and demonstrated this regularity of atom-fixing power in the case of zinc, tin, arsenic and antimony. A serious obstacle, however, in the way of determining the true atomicities of the elements was the general employment of the old so-called equivalent weights, which were by most chemists confounded with the atomic weights. This difficulty was removed by the rectification of the atomic weights, which, though begun by Gerhardt as early as 1842, met for a long time with but little recognition, and was not completed till the subject was taken up by Cannizarro in 1858. The law of atomicity has given to chemistry an exactness which it did not previously possess, and since its discovery and recognition chemical research has moved very much on the lines laid down by this law.

Chemists have been engaged in determining by means of decompositions, the molecular architecture, or *constitution* as it is called, of various compounds, natural and artificial, and in verifying by synthesis the correctness of the views thus arrived at.

It was long supposed that an impassable barrier existed between inorganic and organic substances : that the chemist could make the former in his laboratory, while the latter could only be produced in the living bodies of animals or plants—requiring for their construction not only chemical attractions, but a supposed 'vital force.' It was not until 1828 that Wöhler broke down this barrier by the synthetic production of urea, and since his time this branch of science in the hands of Hofmann, Wurtz, Berthelot, Butlerow, and others, has made great strides. Innumerable natural compounds have thus been produced in the laboratory—ranging from bodies of relatively simple constitution, such as the alcohols and acids of the fatty series, to bodies of such complex molecular structure as alizarin (the principal coloring matter of madder), coumarin (the odoriferous principle of the tonqua bean), vanillin, and indigo. The problem of the natural alkaloids has also been attacked, in some cases with more than partial success. Methylconine, which occurs along with conine in the hemlock, has been recently prepared artificially by Michael and Gundelach, this being the first instance of the synthesis of a natural alkaloid. A proximate synthesis of atropine, the alkaloid of the deadly nightshade, has been accomplished by Ladenburg. It seems further probable that at no distant date the useful alkaloids, such as quinine, may also be synthesised, inasmuch as quinoline, one of the products of the decomposition of quinine, and of some of the allied bases, has recently been prepared by Skraup by a method which admits of its being obtained in any quantity.

Much also has been done in the way of building up compounds the existence of which was predicted by theory. Indeed, the extent to which hitherto undiscovered substances can be predicated is doubtless the greatest triumph achieved by chemists during the past fifty years.

As yet, however, only the statical side of chemistry has been developed. Whilst the physicist has been engaged in tracing, for the gaseous condition at least, the paths of the molecules and calculating their velocities, the chemist, whose business is with the atoms within the molecule, can point to no such scientific conquests. All that he knows concerning the intramolecular atoms, and all that he expresses in his constitutional formulæ is, the particular relation of union in which each of these atoms stands to the others—which of them are directly united (as he expresses it) to other given atoms, and which of them are in indirect union. Of the relative positions in space occupied by these atoms, and of their modes of motion, he is absolutely ignorant. In like manner in a chemical reaction the initial and final conditions of the reacting substances are known, but the intermediate stages—the modes of change—are for the most part unexplained.

The feeling that no number, however great, of successfully solved problems of constitutional chemistry (as at present understood), and no number of syntheses, however brilliant, of natural compounds could raise chemistry above the statical stage—that the solution of the dynamical problem cannot be arrived at by purely chemical means—has led many chemists to approach the subject from the physical side. The results which the physico-chemical methods, as exemplified in the laws already alluded to of Dulong and Petit, Avogadro and Mitscherlich, have yielded in the past, offer the best guarantee of their success in the future. And the advantages of many of the physical methods are obvious. Every purely chemical examination—whether proximate or ultimate—of a compound, presupposes the destruction of the substance under examination: the chemist 'murders to dissect.' But observations on the action of a substance on the rays of light, on the relative volumes occupied by molecular quantities of a substance, on its velocity of transpiration in the liquid or gaseous state—these teach us the habits of the living substance. The rays of light which have threaded their way between the molecules of a body have undergone, in contact with these molecules, various specific and measurable changes, the nature and amount of which are assuredly conditioned by the mass, form, and other properties of the molecules: the plane of polarisation has been caused to rotate; a particular degree of refraction has been imparted; or rays of certain wave-lengths have been removed by absorption; their absence being manifested by bands in the absorption-spectrum of the substance. The volumes occupied by molecular quantities are dependent partly on the size of the molecules and partly on that of the intermolecular spaces.

The duty of the physical chemist is to endeavour to co-ordinate his physical observations with the known constitution of compounds, as already determined by the pure chemist. This endeavour has in various branches of physical chemistry been to some extent successful. Le Bel has found that among organic compounds those only possess action on the plane of polarised light which contain at least one *asymmetric carbon atom*—that is to say, a carbon atom which is united to four *different* atoms or groups of atoms. The researches of Landolt, of Gladstone, and of Brühl on the specific refraction of organic liquids, have shown that from the known constitution of a liquid organic compound it is possible to calculate its specific refraction. Noel Hartley, in an examination of the absorption spectra of organic liquids for the ultra-violet rays, has demonstrated that certain molecular groupings are represented by particular absorption bands, and this line of inquiry has been extended with very interesting results to the ultra red rays by Abney and Festing. It is obvious that these methods may in turn be employed to determine the unknown constitution of substances. The same holds true of the investigations of Kopp with regard to the molecular volumes of liquids at their boiling-points, in which he has established the remarkable fact that some elements always possess the same atomic volume in combination, whereas, in the case of certain other elements, the atomic volume varies in a perfectly definite manner with the mode of combination. This investigation has lately been extended with the best results by Thorpe, and by Ramsay. Thermo-chemistry, also, which for a long time, at least as regards that portion which relates to the heat of formation of compounds, consisted chiefly of a collection of single equations, each containing three unknown quantities, is beginning to be interpreted by Julius Thomsen, whose experienced work in this field is well

MECHANICS. 81

known. Many other methods of physico-chemical research are being successfully prosecuted at the present day, but it would go beyond the bounds of this summary even to enumerate these.

The concordant results obtained by these widely differing methods show that those chemists who have devoted themselves, frequently amid the ridicule of their more practical brethren, to ascertaining by purely chemical methods the constitution of compounds, have not laboured in vain. But the future doubtless belongs to physical chemistry.

In connection with the rectification of the atomic weights it may be mentioned that a so-called natural system of the elements has been introduced by Mendelejeff (1869), in which the properties of the elements appear as a periodic function of their atomic weights. By the aid of this system it has been possible to predict the properties and atomic weights of undiscovered elements, and in the case of known elements so determine many atomic weights which had not been fixed by any of the usual methods. Several of these predictions have been verified in a remarkable manner. A periodicity in the atomic weights of elements belonging to the same class had been pointed out by Newlands about four years before the publication of Mendelejeff's memoir.

In mechanical science the progress has not been less remarkable than in other branches. Indeed to the improvements in mechanics we owe no small part of our advance in practical civilisation, and of the increase of our national prosperity during the last fifty years.

This immense development of mechanical science has been to a great extent a consequence of the new processes which have been adopted in the manufacture of iron, for the following data with reference to which I am mainly indebted to Captain Douglas Galton and Mr. Rendel. About 1830, Neilson introduced the Hot Blast in the smelting of iron. At first a temperature of 600° or 700° Fahrenheit was obtained, but Cowper subsequently applied Siemens' regenerative furnace for heating the blast, chiefly by means of fumes from the black furnace,

G

which were formerly wasted; and the temperature now practically in use is as much as 1,400°, or even more: the result is a very great economy of fuel and an increase of the output. For instance, in 1830, a blast furnace with the cold blast would probably produce 130 tons per week, whereas now, 600 tons a week are readily obtained.

Bessemer, by his brilliant discovery, which he first brought before the British Association at Cheltenham in 1856, showed that Iron and Steel could be produced by forcing currents of atmospheric air through fluid pig metal, thus avoiding for the first time the intermediate process of puddling iron, and converting it by cementation into steel. Similarly by Siemens' regenerative furnace, the pig metal and iron ore is converted directly into steel, especially mild steel for shipbuilding and boilers; and Whitworth, by his fluid compression of steel, is enabled to produce steel in the highest condition of density and strength of which the metal is capable. These changes, by which steel can be produced direct from the blast furnace instead of by the more cumbersome processes formerly in use, have been followed by improvements in the manipulation of the metal.

The inventions of Cort and others were known long before 1830, but we were then still without the most powerful tool in the hands of the practical metallurgist, viz., Nasmyth's steam-hammer.

Steel can now be produced as cheaply as iron was formerly; and its substitution for iron as railway material and in shipbuilding, has resulted in increased safety in railway travelling, as well as in economy, from its vastly greater durability. Moreover, the enlarged use of iron and steel, which has resulted from these improvements in its make, has led to the adoption of mechanical means to supersede hand labour in almost every branch of trade and agriculture, by which the power of production has been increased a hundredfold, while at the same time much higher precision has been obtained. Sir Joseph Whitworth has done more than any one else to perfect the machinery of this country by the continued efforts he has made, during nearly half-a-century, to introduce accuracy into the standards of measurement in use in workshops. He tells us that when he first established his works, no two articles could be made accurately alike or with interchangeable parts. He devised a measuring apparatus, by which his workmen in making standard gauges are accustomed to take measurements to the $\frac{1}{20000}$ of an inch.

In its more immediate relation to the objects of this Association, the increased importance of iron and steel has led to numerous scientific investigations into its mechanical properties and into the laws which govern its strength; into the proper distribution of the material in construction; and into the conditions which govern the friction and adhesion of surfaces. The names of Eaton Hodgkinson, Fairbairn, Barlow, Rennie, Scott Russell, Willis, Fleeming Jenkin, and Galton are prominently associated with these inquiries.

The introduction of iron has, moreover, had a vast influence on the works of both the civil and military engineer. Before 1830, Telford had constructed an iron suspension turnpike-road bridge of 560 feet over the Menai Straits; but this bridge was not adapted to the heavy weights of locomotive engines. At the present time, with steel at his command, Mr. Fowler is engaged in carrying out the design for a railway bridge over the Forth, of two spans of 1,700 feet each; that is to say, of nearly one third of a mile in length. In artillery, bronze has given place to wrought iron and steel; the 68-pound shot, which was the heaviest projectile fifty years ago, with its range of about 1,200 yards, is being replaced by a shot of nearly a quarter-ton weight, with a range of nearly five miles; and the armour-plates of ships are daily obtaining new developments.

But it is in railroads, steamers, and the electric telegraph that the progress of mechanical science has most strikingly contributed to the welfare of man.

As regards railways, the Stockton and Darlington Railway was opened in 1825, but the Liverpool and Manchester Railway, perhaps the first truly passenger line, dates from 1830, while the present mileage of railways is over 200,000 miles, costing nearly 4,000,000,000*l.* sterling. It was not until 1838 that the *Sirius* and *Great Western* first steamed across the Atlantic. The steamer, in fact, is an excellent epitome of the progress of the half-century; the paddle has been superseded by the screw; the compound has replaced the simple engine; wood has given place to iron, and iron in its turn to steel. The saving in dead weight, by this improvement alone, is from 10 to 16 per cent. The speed has been increased from 9 knots to 15, or even more. Lastly, the steam-pressure has been increased from less than 5 lbs. to 70 lbs. per square inch, while the consumption of coal has been brought down from 5 or 6 lbs. per horse-power to less than 2. It is a remarkable fact that not only is our British shipping rapidly on the increase, but it is increasing relatively to that of the rest of the world. In 1860 our tonnage was 5,700,000 against 7,200,000; while it may now be placed as 8,500,000 against 8,200,000; so that considerably more than half the whole shipping of the world belongs to this country.

If I say little with reference to economic science and statistics it is because time, not materials, are wanting.

I scarcely think that in the present state of the question I can be accused of wandering into politics if I observe that the establishment of the doctrine of free trade as a scientific truth falls within the period under review.

In education some progress has been made towards a more rational system. When I was a boy, neither science, nor modern languages, nor arithmetic formed any part of the public school system of the country. This is now happily changed. Much, however, still remains to be done. Too little time is still devoted to French and German, and it is much to be regretted that even in some of our best schools they are taught as dead languages. Lastly, with few exceptions, only one or two hours a week on an average are devoted to science. We have, I am sure, none of us any desire to exclude, or discourage, literature. What we ask is that, say, six hours a week each should be devoted to mathematics, modern languages, and science—an arrangement which would still leave twenty hours for Latin and Greek. I admit the difficulties which schoolmasters have to contend with; nevertheless, when we consider what science has done and is doing for us, we cannot but consider that our present system of education is, in the words of the Duke of Devonshire's Commission, 'little less than a national misfortune.'

In Agriculture the changes which have occurred in the period since 1831 have been immense. The last half-century has witnessed the introduction of the modern system of subsoil drainage founded on the experiments of Smith of Deanston. The thrashing and drilling machines were the most advanced forms of machinery in use in 1831. Since then there have been introduced the steam-plough; the mowing-machine; the reaping-machine, which not only cuts the corn but binds it into sheaves; while the steam-engine thrashes out the grain and builds the ricks. Science has thus greatly reduced the actual cost of labour, and yet it has increased the wages of the labourer.

It was to the British Association, at Glasgow, in 1841, that Baron Liebig first communicated his work 'On the Application of Chemistry to Vegetable Physiology,' while we have also from time to time received accounts of the persevering and important experiments which Mr. Lawes, with the assistance of Dr. Gilbert, has now carried on for more than forty years at Rothamsted, and which have given so great an impulse to agriculture by directing attention to the principles of cropping, and by leading to the more philosophical application of manures.

I feel that in quitting Section F so soon, I owe an apology to our fellow-workers in that branch of science, but I doubt not that my shortcomings will be more than

INTERCONNECTION OF THE SCIENCES. 87

made up for by the address of their excellent President, Mr. Grant-Duff, whose appointment to the Governorship of Madras, while occasioning so sad a loss to his friends, will unquestionably prove a great advantage to India, and materially conduce to the progress of science in that country.

Moreover, several other subjects of much importance, which might have been referred to in connection with these latter Sections, I have already dealt with under their own more purely scientific aspect.

Indeed, one very marked feature in modern discovery is the manner in which distinct branches of science have thrown, and are throwing, light on one another. Thus the study of geographical distribution of living beings, to the knowledge of which our late general secretary, Mr. Sclater, has so greatly contributed, has done much to illustrate ancient geography. The existence of high northern forms in the Pyrenees and Alps indicates the existence of a period of cold when Arctic species occupied the whole of habitable Europe. Wallace's line—as it has been justly named after that distinguished naturalist— points to the very ancient separation between the Malayan and Australian regions; and the study of corals has thrown light upon the nature and significance of atolls and barrier-reefs.

In studying the antiquity of man, the archæologist has to invoke the aid of the chemist, the geologist, the physicist, and the mathematician. The recent progress in astronomy is greatly due to physics and chemistry. In geology the composition of rocks is a question of chemistry and physics; the determination of the boundaries of the different formations falls within the limits of geography; while palæontology is the biology of the past.

And now I must conclude. I fear I ought to apologise to you for keeping you so long, but still more strongly do I wish to express my regret that there are almost innumerable researches of great interest and importance which fall within the last fifty years (many even among those with which our Association has been connected) to which I have found it impossible to refer. Such for instance are, in biology alone, Owen's memorable report on the homologies of the vertebrate skeleton, Carpenter's laborious researches on the microscopic structure of shells, the reports on marine zoology by Allman, Forbes, Jeffreys, Spence Bate, Norman, and others; on Kent's Cavern by Pengelly; those by Duncan on corals; Woodward on crustacea; Carruthers, Williamson, and others on fossil botany, and many more. Indeed no one who has not had occasion to study the progress of science throughout its various departments can have any idea how enormous—how unprecedented—the advance has been.

Though it is difficult, indeed impossible, to measure exactly the extent of the influence exercised by this Association, no one can doubt that it has been very considerable. For my own part, I must acknowledge with gratitude how much the interest of my life has been enhanced by the stimulus of our meetings, by the lectures and memoirs to which I have had the advantage of listening, and above all, by the many friendships which I owe to this Association.

Summing up the principal results which have been attained in the last half-century we may mention (over and above the accumulation of facts), the theory of evolution, the antiquity of man, and the far greater antiquity of the world itself; the correlation of physical forces and the conservation of energy; spectrum analysis and its application to celestial physics; the higher algebra and the modern geometry; lastly, the innumerable applications of science to practical life—as, for instance, in photography, the locomotive engine, the electric telegraph, the spectroscope, and most recently the electric light and the telephone.

To Science, again, we owe the idea of progress. The ancients, says Bagehot, 'had no conception of progress; they did not so much as reject the idea; they did not even entertain it.' It is not, I think, going too far to say that the true test of the civilisation of a nation must now be measured by its progress in science. It is often said, however, that, great and unexpected as the recent discoveries have been, there are certain ultimate problems which must ever remain unsolved. For my part, I would prefer to abstain from laying down any such limitations. When Park asked the Arabs what became of the sun at night, and whether the sun was always the same, or new each day, they replied that such a question was childish, and entirely beyond the reach of human investigation. I have already mentioned that, even as lately as 1842, so high an authority as Comte treated as obviously impossible and hopeless any attempt to determine the chemical composition of the heavenly bodies. Doubtless there are questions the solution of which we do not as yet see our way even to attempt; nevertheless the experience of the past warns us not to limit the possibilities of the future.

But however this may be, though the progress made has been so rapid, and though no similar period in the world's history has been nearly so prolific of great results, yet, on the other hand, the prospects of the future were never more encouraging. We must not, indeed, shut our eyes to the possibility of failure : the temptation to military ambition ; the tendency to over-interference by the State ; the spirit of anarchy and socialism ; these and other elements of danger may mar the fair prospects of the future. That they will succeed in doing so, I cannot believe. I cannot but feel confident that fifty years hence, when perhaps the city of York may renew its hospitable invitation, my successor in this chair—more competent, I trust, than I have been to do justice to so grand a theme—will have to record a series of discoveries even more unexpected and more brilliant than those which I have, I fear so imperfectly, attempted to bring before you this evening ; for assuredly one great lesson of science is, how little we yet know, and how much we have still to learn.

MESSRS. MACMILLAN & CO.'S PUBLICATIONS.

BY THE SAME AUTHOR.

SCIENTIFIC LECTURES.

With Illustrations. 8vo. 8s. 6d.

CONTENTS:—On Flowers and Insects—On Plants and Insects—On the Habits of Ants—Introduction to the Study of Prehistoric Archæology, &c.

'We can heartily commend his volume as a whole to every one who wishes to obtain a condensed account of its subjects, set forth in the most simple, easy, and lively manner.'—ATHENÆUM.

POLITICAL AND EDUCATIONAL ADDRESSES.

8vo. 8s. 6d.

CONTENTS:—On the Imperial Policy of Great Britain—On the Bank Act of 1844—On the Present System of Public School Education, 1876—On Our Present System of Elementary Education—On the Income Tax—On the National Debt—On the Declaration of Paris—Marine Insurances—On the Preservation of Our Ancient National Monuments—Egypt.

'Will repay the careful attention of readers who desire to be acquainted with the best thoughts of a practical and sagacious mind on the most important topics of public and national interest.'—DAILY NEWS.

THE ORIGIN AND METAMORPHOSES OF INSECTS.

With numerous Illustrations. Second Edition.

Crown 8vo. 3s. 6d. (NATURE SERIES.)

'His little book is of great value, and will be read with interest and profit by all students of Natural History.' WESTMINSTER REVIEW.

ON BRITISH WILD FLOWERS CONSIDERED IN THEIR RELATION TO INSECTS.

With numerous Illustrations. Second Edition.

Crown 8vo. 4s. 6d. (NATURE SERIES.)

'All lovers of Nature must feel grateful to Sir John Lubbock for his learned and suggestive little book, which cannot fail to draw attention to a field of study so new and fascinating.'—PALL MALL GAZETTE.

MACMILLAN & CO., London, W.C.

MESSRS. MACMILLAN & CO.'S PUBLICATIONS.

BY PROFESSOR HUXLEY, LL.D., F.R.S.

LAY SERMONS, ADDRESSES, AND REVIEWS. Sixth Edition. 8vo. 7s. 6d.

CRITIQUES AND ADDRESSES. 8vo. 10s. 6d.

PHYSIOGRAPHY. An Introduction to the Study of Nature. With Coloured Plates and Woodcuts. New Edition. Crown 8vo. 6s.

AMERICAN ADDRESSES. With a Lecture on the Study of Biology. 8vo. 6s. 6d.

SCIENCE AND CULTURE, and other Essays. Demy 8vo. 10s. 6d.

BY W. SPOTTISWOODE, LL.D., P.R.S.

POLARISATION OF LIGHT. With numerous Illustrations. Crown 8vo. 3s. 6d. (NATURE SERIES.)

BY PROF. H. E. ROSCOE, F.R.S., AND PROF. C. SCHORLEMMER, F.R.S.

A TREATISE ON CHEMISTRY. With Illustrations. 8vo.
Volumes I. and II. INORGANIC CHEMISTRY. Vol. I. THE NON-METALLIC ELEMENTS. With a Portrait of DALTON. Engraved by C. H. JEENS. 21s. Vol. II. Part I. METALS. 18s. Vol. II. Part II. METALS. 18s.

VOL. III. PART I. ORGANIC CHEMISTRY. 21s.
[PART II. in the press.

BY PROFESSOR H. BALFOUR, F.R.S.

COMPARATIVE EMBRYOLOGY: a Treatise on. With Illustrations. 2 vols. 8vo. Vol. I. 18s. Vol. II. 21s.

BY PROFESSOR MICHAEL FOSTER, F.R.S.

A TEXT-BOOK OF PHYSIOLOGY. With Illustrations. Third Edition, revised. 8vo. 21s.

BY PROFESSOR CARL GEGENBAUR.

ELEMENTS OF COMPARATIVE ANATOMY. By Professor CARL GEGENBAUR. A Translation by F. JEFFREY BELL, B.A., Revised with Preface by Professor E. RAY LANKESTER, F.R.S. With numerous Illustrations. Medium 8vo. 21s.

BY PROFESSOR A. GEIKIE, F.R.S.

ELEMENTARY LESSONS IN PHYSICAL GEOGRAPHY. With numerous Illustrations. Fcp. 8vo. 4s. 6d. QUESTIONS ON THE SAME. 1s. 6d.

OUTLINES OF FIELD GEOLOGY. With numerous Illustrations. Crown 8vo. 3s. 6d.

TEXT-BOOK OF GEOLOGY. With Illustrations. Medium 8vo.
[Immediately.

MACMILLAN & CO., London, W.C.

www.ingramcontent.com/pod-product-compliance
Lightning Source LLC
Chambersburg PA
CBHW032244080426
42735CB00008B/990